Advanced Data Converters

Need to get up to speed quickly on the latest advances in high-performance data converters? Want help choosing the best architecture for your application? With everything you need to know about the key new converter architectures, this guide is for you.

It presents basic principles, circuit and system design techniques, and associated trade-offs, doing away with lengthy mathematical proofs and providing intuitive descriptions upfront. Everything is covered, from time-to-digital converters to comparator-based/zero-crossing ADCs, and each topic is introduced with a short summary of the essential basics.

Practical examples describing actual chips, together with extensive comparisons between architectural or circuit options, ease architecture selection and help you cut design time and engineering risk. Trade-offs, advantages, and disadvantages of each option are put into perspective with a discussion of future trends, showing where this field is heading, what is driving it, and what the most important unanswered questions are.

Gabriele Manganaro is currently an Engineering Director in High Speed Data Conversion at Analog Devices, Inc. He is a Fellow of the IET and has extensive industrial design experience, having previously held positions at National Semiconductor, Engim, Inc., and Texas Instruments.

Advanced Data Converters

GABRIELE MANGANARO

Analog Devices, Inc.

CAMBRIDGE UNIVERSITY PRESS
Cambridge, New York, Melbourne, Madrid, Cape Town,
Singapore, São Paulo, Delhi, Tokyo, Mexico City

Cambridge University Press
The Edinburgh Building, Cambridge CB2 8RU, UK

Published in the United States of America by Cambridge University Press, New York

www.cambridge.org
Information on this title: www.cambridge.org/9781107005570

First published 2012

Printed in the United Kingdom at the University Press, Cambridge

A catalog record for this publication is available from the British Library

Library of Congress Cataloging in Publication data
Manganaro, G. (Gabriele), 1969–
Advanced data converters / Gabriele Manganaro.
  p.   cm.
Includes bibliographical references and index.
ISBN 978-1-107-00557-0
1. Analog-to-digital converters.   2. Digital-to-analog converters.   I. Title.
TK7887.6.M286   2012
621.39′814 – dc23   2011033569

ISBN 978 1 107 00557 0 Hardback

For Nancy, Alessandra, and Umberto

Contents

Preface

Advances in semiconductor technology together with increased demand for complex IC solutions in wireless, consumer, health care or industrial applications have driven significant innovation in data converters during the last decade. While some very good introductory books on the traditional converters and related fundamental concepts are readily available, a coverage of modern advanced analog-to-digital (A/D) and digital-to-analog (D/A) converters and emerging ones is missing. Having to deal with several hundred academic papers and patents as well as tens of specialized monographs aimed at an audience of experts can be frightening and downright disorienting to many practitioners (including IC designers, system designers, and users of data converters) or graduate students attempting to gain insight into one of today's hottest IC areas. This book attempts to cover some of this knowledge gap, offering a bird's-eye view of the relevant principles, the competing requirements, architectures, and circuit techniques, and, perhaps, providing a vision for future developments in this field. It explains the motivations, ideas, and trends associated with the latest and most attention-worthy topics in data conversion of the last ten years or so.

It is assumed that the typical reader has mastered the basics of both analog/mixed-signal design and data conversion. Each subject is introduced by a short summary of the key concepts at its base and complemented by several references guiding readers who might want to look more deeply at each topic.

Acknowledgments

The author wishes to express his appreciation to the many people who, in different ways, guided him and influenced him during his professional journey in IC design and data conversion in particular. These include colleagues and friends at Texas Instruments, Engim, National Semiconductor, and Analog Devices as well as the Data Conversion technical subcommittee of the IEEE ISSCC and those on the editorial board of *IEEE Transactions on Circuits and Systems*.

I would also like to thank Professor Boris Murmann of Stanford University (USA) for useful suggestions on an initial draft of the book, Professor Franco Maloberti of the University of Pavia (Italy), and David Robertson of Analog Devices for support and encouragement. A note of appreciation also goes to the staff at Cambridge University Press, in particular to Dr Julie Lancashire, Miss Sarah Finlay, and Miss Sarah Matthews.

Last but not least, I am extremely grateful to my wife Nancy, my daughter Alessandra and son Umberto, and all the rest of my family for their infinite patience and support.

1 The need for data converters

Modern electronic systems process and store information digitally. However, due to the analog nature of the world, conversion from analog to digital and/or from digital to analog is, and always will be, inevitable and performed by data converters: Analog-to-digital converters (ADCs) and digital-to-analog converters (DACs).

Depending on a variety of factors, including technical specifications, system partitioning, and market needs, data converters can be either integrated on the same chip, or on the same package, together with other analog or digital blocks, or they can be stand-alone. Most stand-alone data converters are part of the $13–$16 billion standard analog market which also includes amplifiers, comparators, and interface and power management devices. In fact, data converters constitute 16% of this market, but they are growing at a faster rate than the other components, second only to power management devices [1, 2], with unit shipments going from about 2.9 billion units in 2010 to an estimated volume of about 4.7 billion units by 2015. That does not account for the embedded data converters [3] integrated together with digital signal processors (DSPs) in a wide variety of applications ranging from consumer electronics (e.g. audio devices, cell phones, imaging devices, DVD and multi-media players etc.) to automotive (e.g. embedded controllers), process control, and instrumentation.

This chapter will briefly outline some of the applications of data converters, illustrating the pervasiveness and variety of ADCs and DACs in modern predominantly digital signal processing systems. Moreover, as integrated digital systems continue to develop in line with CMOS technology scaling, and systems-on-a-chip (SoC) become ubiquitous, it is inevitable that data converters will need to cope with the needs of the applications and face the challenges and opportunities offered by ever shrinking process technologies. A second part of this chapter will be devoted to a summary of some of the issues, trade-offs, and solutions that data-converter designers face when designing analog circuits in the context of a technology road-map primarily driven by digital circuits needs. The chapter will also point to some of the alternatives to marching down the route indicated by Moore's law.

The concepts summarized in this chapter provide a context and motivation for the advanced data converters covered later in this book and will help us to develop a vision and spot the trends in this exciting field.

1.1   The digital revolution in an analog world: some examples of applications of data converters

As pointed out above, a digital signal processing (DSP)-based system needs ADCs and DACs in order for it to be able to exchange information or to control processes in the physical analog world. The number of applications of signal processing and control is extremely large and constantly expanding, and the aim of this section is to give the reader, by means of a few examples, a sense of the variety of the requirements and how this diversity leads to a wide range of different data conversion architectures, each one aiming to provide the optimal conversion solution to a different set of technical needs and specifications.

As stated before, it is assumed that the reader possesses a good understanding of the traditional A/D and D/A architectures and the fundamentals of data conversion and sampling theory.

1.1.1   Sensing applications

We will begin with precision measurements and sensor signal conditioning applications [4], including the digitization of signals originating from a wide variety of sensors[1] such as pressure, temperature, gas, speed, acceleration, and light-intensity sensors. An important part of this application space concerns temperature sensors used to monitor printed-circuit boards (PCBs) and microprocessor die temperature.

The applications we are referring to are generally characterized by relatively low-bandwidth signals ranging from DC to kHz at most. On the other hand, such signals can require high-precision digitization and/or may coexist with undesired signals due to, for example, electromagnetic interference, noise etc. Moreover, the electric variables transducing the sensed physical variables vary quite a bit. For instance, strain gages, flow meters, and pressure sensors are often variable resistors. Other sensors, such as some position sensors, behave as variable capacitors and so on. Depending on the interface circuitry between the actual ADC and the sensor, the input to the ADC may be a voltage or a current. Examples of interfaces between the sensor and the ADC include a resistor or a capacitor bridge, a force-sense system etc. There are many other important considerations, which go beyond the scope of this book. However, the common elements of many of these applications can include the following:

- very high nominal output resolutions up to even 24 bits with no missing code;
- very low noise level, for example up to 19 bits of noise-free code resolution;
- some level of on-chip digital filtering; for instance allowing excellent 50 Hz/60 Hz power supply rejection;
- low offset;
- low gain error;
- low temperature dependence;
- low pin count package.

[1] Possibly after signal conditioning including analog filtering, amplification, level shifting etc.

Table 1.1. Typical specifications for digital audio converters [7]

Application	THD+N (dB)	f_s (kS/s)
Telecom	60–70	8
FM stereo	60–70	32
Speech analysis	70–80	8–48
Computer audio	80–90	48
Stereo CD, DAT, etc.	>100	44.1, 48, 88.2, 96
DVD audio	>100	48, 96, 192

The converters that have traditionally been used in applications like these are the so-called "serial" ADCs such as *single-slope* and *dual-slope* ADCs [5]. The principle of operation of such ADCs consists of comparing an internally generated accurate voltage ramp with the input signal (assuming that this does not appreciably vary while the conversion is ongoing) and counting how many clock cycles elapse between the start of the ramp and the instant at which the ramp reaches the input signal. These integrator-type converters must rely on very linear ramp generators, stable clock sources etc., and, as will become clear from the second part of this chapter, have fundamental issues with the low-voltage supplies of advanced CMOS processes.

For these and other reasons, in recent years, high-resolution $\Delta\Sigma$ ADCs and incremental ADCs[2] have gained considerable popularity in these applications. Moreover, these types of converters are easily implemented in CMOS, and greater flexibility is obtained by adding many on-chip digital programmable functions including filtering, variable throughput rate, programmable gain, calibration modes etc. while requiring few or no external components. In addition to that, since these are oversampled converters, the requirements on the anti-aliasing filters are considerably relaxed due to the large difference between the signal bandwidth and the Nyquist frequency, making the entire front end simpler, smaller, and cheaper.

1.1.2 Digital audio

This market includes audio circuits in portable applications such as wireless handsets or CD and MP3 players as well as in fixed applications such as high-fidelity (HiFi) systems and professional/studio audio systems [7].

In the case of speech digitization (e.g. in cell phones) the standard sampling frequency is $f_s = 8$ kS/s, and 16-bit linear $\Delta\Sigma$ ADCs and DACs have replaced the traditional logarithmic converters used in the early systems. In fact, DSP-based speech compression algorithms reduce the overall data rate to acceptable levels. More than resolution, total harmonic distortion plus noise (THD+N) is an important parameter and Table 1.1 shows typical values of the THD+N and sample frequency f_s for various digital audio applications [7]. $\Delta\Sigma$ ADCs and DACs also dominate in the professional audio area, and,

---

[2] An incremental ADC [6] is a particular type of $\Delta\Sigma$ ADC that is re-set after the conversion of each sample. Better coverage of this type of ADC is deferred to subsequent chapters.

once again, offer the advantage of simplifying the anti-aliasing and reconstruction filters owing to oversampling. Furthermore, oversampled DACs have the additional advantage of suffering less from the $\sin(x)/x$ spectral shaping issue[3] thanks to both oversampling and the use of digital interpolators/filters preceding the DAC itself.

There are also other important specifications that are very specific to audio systems, such as containing the power of specific harmonics, but that is well beyond the scope of this chapter.

1.1.3   Wireless communication infrastructure

Mobile telephony, despite going through market cycles, has overall seen a significant steady increase in demand for mixed-signal products, including data converters. Basic cell phones have been adopted widely in Western countries as well as in Japan and Korea, and they are quickly penetrating populous countries and emerging markets such as China and India as well as many South American countries [8], hence driving increasing production. Furthermore, compounding this increasing pool of users, the emergence of smart phones with their associated higher demand for bandwidth and services means that the wireless communication infrastructure is going to require a significant expansion that is somewhat similar to what has been experienced by the wireline infrastructure as a result of the boom in wired internet access [9].

In wireless communication infrastructure applications, the data converters are critical blocks of base transceiver stations (BTSs), namely the wideband transceivers that handle the communication with the many handsets active in the corresponding phone cell [10, 11, 12]. Such converters, in order to acquire or synthesize multiple user channels, at present have signal bandwidths of the order of a few tens of MHz and are steadily headed toward the 100 MHz mark. BTS systems can be required to be able to handle multiple communication standards such as GSM, CDMA, LTE, wideband CDMA, and WiMax (or IEEE 802.16) [13]. This means that higher flexibility, integration (hence also reduced cost and power consumption per digital function), and lower overall costs should be achieved by placing the ADCs and DACs as close as possible to the antenna [14], hence the shifting of some of the less flexible, expensive, high-count parts, which are traditionally analog functions, to the digital domain. Some of these analog functions include filtering, using surface acoustic wave (SAW) filters, ceramic filters, or crystal filters, with tight tolerance and matching requirements for frequency-sensitive components such as inductors and capacitors, mixing down/up-converters, oscillators etc. Because of that there is an increasing demand, where sensible, to move from zero-intermediate-frequency (IF)/low-IF schemes whereby the ADCs or the DACs convert baseband signals to high-IF digitizing/synthesizing schemes whereby the IF-to-baseband frequency down-conversion/up-conversion stage (which may be a quadrature scheme to take care of image rejection), possibly with some filtering, is moved from the analog to the digital domain. Ideally, the final goal would be to put the ADC right after the

---

[3] Because the desired output signal resides at a frequency that is much lower than the first frequency null of $\sin(x)/x$ due to the much higher f_s.

low-noise amplifier (LNA), aside, perhaps, for some filtering, and in front of a receive signal processor (RSP), and the DAC right before the power amplifier (PA), aside, again, for some filtering, and following a transmit signal processor (TSP), fully realizing what is commonly referred to as software radio (SR) [10, 15, 16, 17].

Some of the issues stemming from bringing the converters toward higher-frequency stages of the transceiver are related to its sensitivity, selectivity, out-of-band (image) rejection, ability to handle undesired powerful "blockers" etc. Moreover, converting such high-frequency signals, with even increased dynamic specifications as the analog–digital interface is moved closer to the antenna,[4] is a formidable challenge and quickly results in prohibitive power consumption levels.

The radio specifications consist of the noise level, measured in terms of its noise spectral density (NSD), and the linearity with which signals are processed, measured typically by means of two-tone or multi-tone intermodulation distortion. Such specifications, once the bandwidth has been specified, result in converter specification. For ADCs in IF digitizing applications these include SNDRs in the range between the low 70s (dB) and the mid-to-high 80s (dB). Often these specifications can be traded, at least to some extent, with the PGA gain placed in front of the ADC and the selectivity of the preceding filters. These converters are often "SNR-limited" (namely, the RMS noise is comparable to or greater than the LSB size) and have very low distortion. The latter is often quantified by the spurious-free dynamic range (SFDR) for wideband cases and by the intermodulation distortion (IMD) for narrowband cases.

The sampling clock phase noise (PN) is also a critical specification due to its negative effects on the digitized noise floor (particularly at frequencies close to that of the digitized signal) as a result of sampling aperture uncertainty, and it becomes increasingly worse as a function of the increasing digitized signal frequency.

For converters used in handsets, in addition to the particular engineering performance specifications enabling the communication standard, some of the key specifications result from the need for products (1) to be small in size, (2) to be inexpensive, and (3) to have a long stand-by time (long battery life). This is achieved through employing higher integration levels (as pointed out before, digitizing earlier in the processing chain reduces the number of passive components and the overall part count) while, at the same time, striving to improve power efficiency [18].

On the other hand, in BTS applications, converters capable of meeting the associated engineering performance specifications can be very power hungry, demanding up to 1 W per converter core or more. However, BTSs are large and power-hungry fixed systems, and, although about 30% of the operation cost is electric power consumption, the contribution of the converters to the overall power consumed by the transceiver is not very significant (as opposed to, for example, the PA or some of the driving amplifiers that make up for the loss in filters and/or drive the inputs of ADCs). Therefore obtaining the desired noise and distortion performance is typically the most important goal, and higher power consumption is an acceptable cost to pay for it. Also, their size

---

[4] Which means that the signal now includes a lot of undesired content that would otherwise have been filtered out and that now interferes with the desired content.

is not necessarily a very critical factor (and, although it impacts production costs, the average selling prices are typically large enough as a result of the level of performance delivered by these parts) since they take up a relatively small amount of space compared with some of the filters and the system as a whole, and, again, these are not mobile systems.

Somewhat conceptually similar to BTS applications are wireless LAN (sometimes referred to as *WiFi* or, more generally, as the IEEE 802.11 standard) applications, where the *access point* (which is homologous to the BTS in this application) is required to handle multiple "clients" (e.g. laptops, workstations etc.) in a relatively small environment such as an office space, a public library, a train or an airplane, a coffee shop etc.

Converters for such applications are somewhat similar to those just cited in the BTS case, but with considerably more relaxed dynamic specifications. It should be noted that there is at present a trend toward convergence between WiFi, its larger-range sibling WiMAX, and cell phone BTSs (in fact, in the case of so-called pico-cells, a cell is extremely constrained in physical space). This is, again, something that could be made to become more and more real by taking advantage of the flexibility offered by SRs. The catch, of course, is in the details of its implementation.

A slightly different category of wireless communication applications is constituted by satellite communication and military communication applications. In the former case, while the dynamic specifications tend to be more relaxed than in the BTS case, the bandwidth of the signal is considerably wider (converters with $f_s > 1$ GS/s are not unusual). A peculiar requirement in space applications, however, is the ability of the converter to be radiation-tolerant since the chip is meant to be part of a satellite system in space and therefore is going to be exposed to radiation and high-energy particles. Such exposure, if proper design measures/techniques are not employed, can rapidly lead to latch-up (transient or permanently destructive) internal to the chip being induced by radiation and bombardment of ions and particles. Packaging is another challenge, and special packages are often required.

In military applications, similarly to the space applications, a common requirement is a very high sampling frequency and, again, relatively relaxed SNDR.[5] However, here, the environmental conditions (e.g. the temperature range) under which one needs to guarantee the converter's performance are extreme.

1.1.4   Health care and life sciences

Electronics for health care is another application area that is projected to grow considerably in the near future. This is a very diverse application field insofar as it includes new-generation implementations of "traditional" equipment such as ultrasound scanners, CAT scanners, X-ray machines, blood pressure monitors etc. as well as newer types of devices such as diagnostic devices taking advantage of ultrawideband wireless

---

[5] In reality it is not that it is relaxed but rather that it is not yet feasible to have single ADCs with more than 12–14 ENOB *and* a sample rate in excess of 3–5 GSPS! So the system designer will need to make do with what is available.

technology for breast cancer screening or pocket-sized CPR devices using micro-electro-mechanical accelerometers and able to perform real-time measurements of the rate and depth of CPR chest compressions [19]. The variety of system implementations (and the corresponding bandwidth and dynamic range of the signals being processed) is huge.

For example, ultrasound machines take advantage of phased array signal processing to be able to estimate spatial information together with density information. Correspondingly, large arrays (16, 32, 64, and more) of high-voltage DACs are often required as part of the synthesizer for the ultrasound pulse launched toward the organs being scanned. Likewise, large arrays of ADCs are critical components of the receive path digitizing the returning ultrasound pulse. At present, for example, medium-resolution ultrasound machines use arrays of ADCs with SNR in the low 70s (dB) and input signal bandwidth of the order of tens of MHz [20]. In such applications, due to the large number of DACs/ADCs required, together with the dynamic specifications (SNR, THD, etc.), it is also important that the individual converters have low power consumption and low area in order to allow the highest possible level of integration and, increasingly, low weight and long battery life in order to also enable portable devices. Both pipelined and $\Delta\Sigma$ ADCs are used in this type of application.

A completely different case is, for example, the one where miniaturized arrays for gene-based tests are used in DNA testing and the signals have sub-kHz bandwidth and a complex chip, where the A/D (basically an incremental A/D) is only one of the blocks, using only microwatts of power [21]. Another case is that of a chip to monitor vital signs such as temperature, heart rate, and electrocardiogram (ECG/EKG), where the integrated 10 bit/500 Hz $\Delta\Sigma$ ADC works at 1 V supply and requires 20 µW [22].

1.2 Challenges and opportunities offered by recent technology advancements

Progress in the development of silicon technology continues to allow deeper levels of CMOS transistor scaling and integration [23, 24]. Moore's law [25] has continued to predict the rate of integration fairly well for decades and, actually, the rate has even increased in recent years [23, 26]. That has both been driven by system needs and has made new applications possible in a somewhat symbiotic fashion. Despite the various types of physical and technological challenges, CMOS scaling is going to continue advancing for many years to come thanks to innovation in the area of materials and device design. In addition to higher levels of integration and reduced cost of functionality, scaling benefits digital circuits in terms of power efficiency and speed.

Unfortunately, however, analog circuits such as ADCs and DACs do not necessarily benefit from CMOS scaling in the same way as that in which digital circuits do [27, 28, 29, 30, 31], as will be briefly summarized in Section 1.2.1. However, as discussed above, to continue to exploit the system advantages offered by digital functionality, it is necessary for data converters in many application spaces to deal with the challenges of ever shrinking CMOS processes.

Table 1.2. Examples of voltages associated with CMOS technology nodes

Technology node (nm)	Analog V_{DD}	Digital V_{DD}	V_T
250	3.3	2.5	0.60
180	3.3	1.8	0.45
130	3.0	1.2	0.40
90	2.5	1.0	0.30
65	1.8	1.0	0.30
45	1.2	0.8	0.25

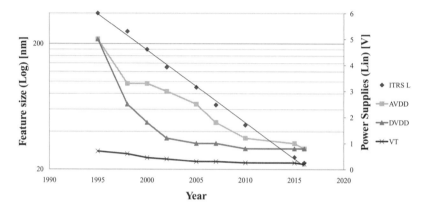

Figure 1.1 CMOS process scaling: feature length, associated supplies, and threshold voltages.

Yet, in some other cases, alternatives to full integration in deep sub-micron CMOS processes do exist. This includes using BiCMOS processes, multi-chip modules, or thoughtfully defined chip sets as discussed in Section 1.2.2.

1.2.1   Main issues with scaling in analog circuits

A first and fundamental issue is the reduction in power supply V_{DD} as the thickness of the thin oxide of the MOS gates T_{ox} is decreased.[6] Table 1.2 and Fig. 1.1 show examples of supplies and MOS threshold voltages for various CMOS process nodes. The available signal swing is reduced accordingly because of the reduction in voltage headroom in critical circuits such as operational amplifiers, comparators etc. Unfortunately, the noise floor does not scale down while the threshold voltages do not scale at the same rate as that of the supplies.[7] Owing to the lower supply, the common-mode range of differential circuits is also reduced, aggravating the swing limitations further.

---

[6] Assuming a SiO_2 dielectric, the oxide thickness T_{ox} scales proportionally to the minimum feature size L_{min} as $T_{ox} = \lambda L_{min}$, with $\lambda \sim 0.03$ [29].

[7] Lowering V_T leads to excess subthreshold current $I_{OFF} \sim \exp[-V_T q_e/(n k_B T)]$ when the channel is not formed. However, some processes provide different choices of devices with different threshold voltage levels, including, sometimes, zero-threshold devices.

Although, as shown in Table 1.2, mixed-signal CMOS processes offer "analog" MOS transistors with a higher nominal supply than that of the "native" digital MOS devices, the former have higher gate oxide thicknesses (hence, these are sometimes referred to as "dual gate oxide," DGO, devices) and longer minimum channel lengths than the latter. In other words, for a given process node, the designer can take full advantage of the scaled digital circuits while being able to design analog circuits on the same chip with devices from roughly one generation (or more) behind. A considerably more useful process option is the one of using "asymmetric" thin-oxide *high-performance analog* (HPA) transistors, with which a specially designed drain implant area leads to higher output impedance at relatively short channel length with clear benefits for op-amp design [32].

As the voltage headroom is reduced, the use of (especially multiple) cascoding to implement high DC gain in operational amplifiers becomes increasingly difficult. To cope with that, designers replaced multiple levels of cascoding with regulated/boosted cascoding, or ultimately resorted to using multi-stage (e.g. Miller or nested-Miller and then feed-foward compensated amplifiers) amplifiers [33, 34, 35, 36, 37, 38, 39, 40]. However, some of these architectural choices can either lead to the aggravation of introducing multiple poles at lower frequency than in a traditional single-stage cascoded amplifier, therefore ending up with actual lower closed-loop bandwidth designs, or are difficult to use in traditional switch capacitor circuits due to their transient response.

CMOS switches (sometimes referred to as "transmission gates") also suffer greatly from the supply reduction (see Fig. 1.2(c)). The on-resistances (R_{onN} and R_{onP}) of the parallel-connected NMOS and PMOS transistors constituting the transmission gate strongly depend on their respective transistor's overdrive voltage (e.g. $V_{OD} = V_{GS} - V_{TN}$ for the NMOS transistor) (see Fig. 1.2(d)). If the gates of these two transistors are driven to the supply levels (e.g. $V_G = V_{DD}$ to turn on the NMOS transistor) while the passing input signal V_{in}, tied to the source terminal ($V_S = V_{in}$), is at a voltage level that is becoming too close to V_T volts away from the gate voltage level, then the overdrive voltage (e.g. $V_{OD} = V_{DD} - V_{in} - V_{TN}$ to turn on the NMOS switch; see Fig. 1.2(a)) may become too close to zero or even negative. As a result, the corresponding MOS switch will not be able to have a sufficiently small on-resistance or it could be completely switched off while it was desired to be turned on (see Fig. 1.2(b)). So, while a passing signal close to either V_{DD} (leading to low R_{onP} in the PMOS switch) or ground (leading to low R_{onN} in the NMOS switch) will cause at least one of the two parallel transistors to be fully turned on with a large overdrive, on the other hand, a passing signal at a level around $V_{DD}/2$ may lead to insufficient overdrive voltage on, possibly, both NMOS and PMOS transistors, which could even turn off [33, 34] (see Fig. 1.2(e)). The latter will happen, for example, as the absolute sum of the two threshold voltages $||V_{TN}| + |V_{TP}||$ becomes equal to or larger than V_{DD}.

In order to address the issue of insufficient overdrive voltage in MOS switches, techniques involving driving the gate voltage outside the range set by the supply voltages (i.e. to a potential higher than V_{DD} for the NMOS switch and to a potential lower than ground for the PMOS switch, respectively) have been developed. Generally, two types of techniques are used, known as "boosted supplies" and "boosted switches," respectively. With "boosted supplies" an internal voltage level higher than V_{DD} (or lower

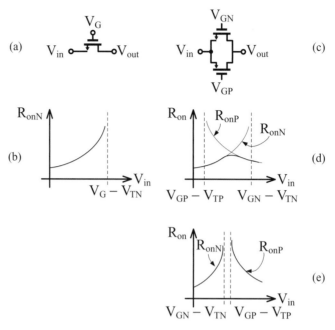

Figure 1.2 MOS switches. (a) An NMOS switch. (b) The on-resistance of the NMOS switch R_{onN} as a function of the input voltage V_{in}; the resistance becomes increasingly higher as the input approaches $V_G - V_{TN}$. (c) A CMOS switch. (d) The on-resistance of the CMOS switch R_{on} as a function of the input voltage V_{in}; since $R_{on} = R_{onN} // R_{onP}$, for sufficiently high overdrive voltages on the NMOS and PMOS transistor, R_{on} is low and relatively constant. (e) The same R_{on}, however, for very low voltage and, hence, insufficient overdrive on the NMOS and PMOS, can become very high or nearly infinite when V_{in} nears the middle of its range; the switch can conduct only when V_{in} is near the supply and ground as the NMOS and PMOS switches separately get sufficient overdrive to turn on.

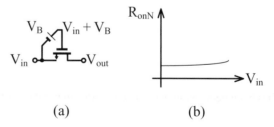

Figure 1.3 (a) An NMOS boosted switch. (b) R_{on} of the boosted switch versus V_{in}; the slight increase of R_{on} for higher V_{in} is due to the "bulk effect," namely the increase of the threshold voltage for increasing V_G.

than ground) is generated (e.g. using charge pumps) to be used to drive the gates of the MOS switches. With "boosted switches" [41], instead, a "floating battery" V_B is generated and then connected across the gate–source of the MOS switch, to turn it on, so that its gate voltage level "rides" V_B volts away from the passing signal as shown in Fig. 1.3. Conceptually, for example, this could be realized by first connecting a capacitor between V_{DD} and ground and then connecting it between the gate and source of the MOS switch to turn it on. The obvious challenge to deal with is the immediate drop in "battery" voltage upon connection to the switch, caused by charge sharing between the boosting capacitor and the MOS gate–source capacitance. Also, various leakage currents will discharge the boosting capacitor over time.

Both approaches have to be implemented with great care because, while the circuits get powered on or off, or even during normal operation, some of the devices may experience stressful or even destructive voltage transients that can lead to undesired behaviors (e.g. accidentally forward biased junctions that were meant to stay reversed biased, such as the drain-to-substrate junction) or their long- and short-term reliability may be affected (e.g. when V_{GS} swings higher than the oxide breakdown voltage). Furthermore, these internally generated voltage sources need low impedance in order to be able to quickly turn on the desired switches (charging their gate capacitances and various other stray capacitances). This could lead to relatively large, power-hungry, and possibly noisy boosting circuitry [33, 34]. So, most often, it will be practical to boost only a few critical switches but handle all others in a more traditional way.

Driving the back-gate (bulk terminal) of the MOS switch to alter its threshold is a very tricky option, especially since this might not always be an available terminal (it depends on the well implementation of the devices) and it is likely to be one with a large capacitance associated with it. This is often not even modeled in the device deck model.

Active device matching is another important matter. This is necessitated by threshold voltage mismatch ΔV_T and the current factor mismatch $\Delta \beta$ (where $\beta = \mu C_{ox} W/L$) with the well-known empirical relations [42, 43]

$$\sigma^2(\Delta V_T) = \frac{A_{VT}^2}{W \cdot L} \tag{1.1}$$

$$\left(\frac{\sigma(\Delta \beta)}{\beta} \right)^2 = \frac{A_{\beta}^2}{W \cdot L} \tag{1.2}$$

Active device mismatch for a given process is then quantified using *Pelgrom's matching parameters* A_{VT} and A_{β}, and, in practice, the threshold voltage mismatch A_{VT} is the dominating factor, for example, in the drain current mismatch between two adjacent (ideally) identical devices. This is, for example, a representative case for input differential pairs in amplifiers and comparators or for matching pairs in current mirrors. Although A_{VT} has been decreasing over finer and finer generations of technology nodes,[8] it has not dropped as rapidly as the supply and signal swing. Therefore, for example, amplifiers' offsets originated by input pair mismatches tend to be relatively larger in comparison with the available swing as subsequent technology nodes are considered [43]. Furthermore, on technologies finer than 180 nm, this model does not properly capture what is observed with small-feature devices [44], and even greater variability is clearly exhibited at 90 nm and for deeper sub-micron processes [24, 45].

As has already been pointed out, besides all the above-cited challenges associated with lowering supplies, due to a number of physical and technological factors, deep

---

[8] In fact, it has been shown that $A_{VT} = \gamma T_{ox}$, where γ is a technology-independent constant [29].

sub-micron MOS devices (beginning even at 180 nm and considerably exacerbating on moving into subsequent nodes) exhibit a large variety of phenomena making the classic MOS drain current square law progressively limited in its ability to model the transistor behavior. Consequently, for example, the classic equations relating small-signal parameters to the quiescent point mutate as well, and can become limited to very specific regions of biasing. In [31] it is shown that on, for example, going from 180 nm to 90 nm

- while the transconductance efficiency g_m/I_D is unaffected
- the device's unity gain frequency f_T improves, and
- the intrinsic gain g_m/g_{ds} degrades.

Device effects becoming increasingly important as feature sizes shrink include the following.

- *Velocity saturation*: carriers in the inverted channel quickly reach the maximum transfer speed, hence limiting the drain current.
- *Gate current leakage*: undesired direct current begins to flow through the thin oxide between the gate and the channel due to tunneling through the oxide. To make things worse, this current is exponentially dependent on the voltage across the gate oxide.
- *Drain-induced barrier lowering* (DIBL): V_T becomes appreciably dependent on the drain potential; this affects the depletion in this region and the impedance looking into the drain and, as technology scales, it becomes dominant over the usual channel-length modulation.

These and other effects are temperature-dependent, and, in deeper and deeper process nodes, can vary appreciably in parametric strength from process run to process run, from wafer to wafer, from die to die, on a die-scale, and even all the way down to circuit block level, taking the old design challenge of dealing with process, voltage, and temperature (PVT) sensitivity to a whole new level [46]. As noted in [24], as CMOS approaches the 25 nm node, stochastic threshold variation caused by dopant implant position in ultra-small inversion regions will give rise to more than 100 mV of threshold variation for minimum-size devices, and, in addition, process proximity effects induced by layout [47], loading effects caused by device density, and gate line-edge roughness will bring additional contributions to the variation [45].

In [29] the implications of the main scaling issues have been related to some classic data converter architectures by analyzing fundamental circuit blocks. It has been shown that circuits in which active device matching is the dominant factor in obtaining a desired dynamic range have the largest power consumption but tend to have decreasing or constant power consumption with shrinking technologies, hence benefitting from scaling. An example of this is flash-type ADCs.

On the other hand, those circuits in which noise dominates the design specifications show an increasing or equal power consumption as technology scales. An example of this is pipelined ADCs.

Power consumption is a meaningful parameter to consider in such comparisons since it can be generally assumed that higher dynamic performance of analog circuits can be ultimately obtained by burning more power [29]. That is also why power efficiency is such an important metric for comparing various designs, and it is often the critical element used in various factors-of-merit (FOMs), which will be discussed later on in this book.

Last but not least, we come to the realization that process technology has outpaced design and simulation capabilities. We are able to integrate larger systems than we are able to effectively simulate, analyze, and test. Simulators have to deal with exponentially increasing transistor count and the device models too are becoming dramatically complex if critical effects must be captured.

Moreover, it traditionally takes multiple process runs for device models to become acceptably accurate for most mixed-signal design needs.[9] However, since subsequent technology nodes are also becoming exponentially more expensive [26], it is often not possible to count on too many iterations for the purpose of modeling.

1.2.2   Alternatives to scaling

During the last thirty years or so, besides mask count, one of the attractivenesses of CMOS processes compared with bipolar processes, for the implementation of data converters, has been the ability to add digital functionality without incurring significant cost, power, or size penalties. For example, this has been crucial to those data converter architectures with large digital sections such as $\Delta\Sigma$ converters, but also to a wide variety of application-specific converter products. Examples of that include the integration, together with PGAs, ADCs, and DACs, of digital filtering functions, multiplexing, encoding/decoding, modulation/demodulation, data compression etc. for which CMOS is definitely the ideal choice. Chips of this kind are designed for important applications in display electronics, wireless communication, instrumentation etc. These are sometimes referred to as "analog front ends" (AFEs) because a complex, often multi-channel, cascade of mixed-signal blocks is integrated with the digital post- or pre-processing functionality.

The reader should be reminded again that, although the fabrication/wafer cost has been rapidly increasing over subsequent generations/technology nodes, the higher level of integration of the digital functions has outpaced that, hence leading to a net drop in digital functionality's cost. On the other hand, as already implied from some of the prior sections, the same cannot always be said for the analog blocks (in fact, the opposite is often true). It is therefore important to determine, from case to case, the respective weights of the digital and analog sections, before selecting a suitable CMOS node [3, 29].

Another strength of CMOS processes over bipolars has been the ease of implementation of switches (they are easy to implement and drive, with low leakage, relatively low on-resistance and low stray capacitances, low area). Yet, analog blocks such as amplifiers and voltage references can get their best performance when taking advantage

---

[9] It would take another whole book to discuss this matter. Certainly, designing data converters with resolution and accuracy well in excess of 16 bits, with models that when "mature" are routinely accurate to 8 bits or so, should give a sense of the design skills necessary if one is to be successful in this field.

of bipolar transistors. A BiCMOS process, if properly utilized, can then virtually offer the best of both worlds to the mixed-signal designer faced with implementing very high performance specifications. Besides, the costs associated with BiCMOS processes are higher but still tolerable when compared with those of the corresponding mixed-signal CMOS nodes, especially when the BiCMOS process at hand is essentially devised as the addition of one or two (typically a high-voltage and low-voltage version) NPN bipolars to an existing CMOS base. Such higher costs are generally fully justified by the higher average selling price (ASP) of such high-performance chips [48, 49].

We should also mention that there are many data converters implemented in compound semiconductor technologies to achieve ultra-high speed. However, in order to bias the devices for optimum speed, large current levels are necessary, and this results in significant power consumption and, consequently, triggers a number of other engineering choices and challenges associated with the packaging and the board, as well as the design of the whole signal chain associated with it. Furthermore, such exotic process technologies allow integration scales that are significantly lower than that of standard CMOS (let alone the most advanced processes) and offer dramatic challenges to digital design. Compound semiconductor converters hence find space in niches such as military, ultra-high-frequency, and some space applications, where costs, power consumption, size, weight etc. are inevitable admission tickets to the game.

In other cases, a single-chip integration might also not necessarily be the best technical and economical choice. A complex system may be, in fact, more practical and economical when implemented either as a chip set or by packaging multiple dies (each using a different process technology) in a single package. The latter is the case of multi-chip modules (MCMs). This is meaningful when, for example, a two-chip design may be smaller, faster, better, or cheaper than a single integrated solution [50].

Note, however, that the question is not only which process technology is best suited for each function, but also includes finding out where the optimal chip boundaries lie when considering the whole system. It might not be as simple as "analog on one chip, digital on another (possibly, memory on a third)" [51, 50] or even "driving amplifier on one chip, data converter on another (possibly, clock synthesizer on a third)." The optimal partitioning must account for the flow of the signal information in various sections of the system, as well as the IP strengths of the chip providers (which, in fact, may be different). For example, considering data exchange between various sections of the system, in order to keep electromagnetic interference (EMI) low, it is desirable to have low-data-rate interfaces exchanging information in/out of the chip, while keeping high-speed busses on-chip. This means that some digital functions are better integrated on the same die with the analog sections. Examples are the integration of the interpolation filters together with communication D/As, the integration of digital filters, decimation filters, down-converters and so on, together with communication A/Ds, etc.

The architectural choice of the die-to-die or part-to-part interfaces is also very critical. For example, for data rates in excess of 200 MHz it is generally preferable to use low-voltage digital signaling (LVDS) because, in comparison with traditional CMOS/rail-to-rail signaling, it offers better signal integrity and better power consumption at higher speed, at the cost of doubling the corresponding pin count. As mentioned before, when dealing with a very-high-data-rate source, it may be preferable to increase the pin count by adopting multiple lower-data-rate multiplexed buffers. In contrast, when one is dealing with a lower-data-rate source and a lower pin count is desired, the data can be serialized at the chip interface and then de-serialized on the receiving chip.

It is apparent that the optimal partitioning involves several technical (design but also testability) considerations but also cost and product-development-time considerations. This is the complex problem of "smart partitioning" [50, 51]. Once again, simulating and modeling this type of system is extremely challenging, especially in the case in which multiple chips (each using a different process technology, possibly on different packages, each one with its own model) are simultaneously being emulated at a relatively low level of abstraction.

1.3 Summary

Data converters are needed and pervasive in all modern electronic applications requiring digital signal processing. A quick snapshot of some important application areas has been provided to give the reader a sense of the broad variety of requirements and specifications involved. Such variety cannot simply be matched by a single "universal" type of ADC or DAC. Different architectures of converters address these diverse needs, as discussed in Chapter 2.

Furthermore, because of the close link between ADCs/DACs and DSPs, CMOS processes and the consequences of their road-map must be dealt with by designers, as discussed in Section 1.2.1. Alternatives to CMOS scaling, however, do exist, as covered in Section 1.2.2.

As IC technology progresses, new challenges and opportunities are offered to the data converter community. This chapter provided a context for the advanced data converters described in the following chapters.

2   A refresher on the basics

This chapter will provide a "refresher" on some of the background topics necessary for the following chapters. It is assumed that the reader is already relatively familiar with the basics of characterizing noise and distortion, with the principles behind the classic data converter architectures, and with some rudimentary idea of calibration. These are, in fact, topics that are much more extensively covered in other books and publications [5, 34, 52, 53, 54]. The intent here is to recall them and to build upon them by expanding the topic a bit more than is usually done in analog design textbooks, to prepare the ground for the more advanced topics covered in the following chapters.

With that in mind, the presentation style has been intentionally kept somewhat informal and, at times, also deliberately "high level" (others will say "simplistic") to avoid digressing into topics that would require a very large tome if they were to be covered satisfactorily.

2.1   Mapping needs to performance metrics

Traditionally, the main specifications characterizing ADCs and DACs are the resolution (or number of bits) n and sample frequency f_s. However, as we will see in the following, other parameters are much more meaningful for characterizing the performance of data converters, depending on the context and application.

To begin with, let us remember the difference between *accuracy* and *precision*. Accuracy refers to the degree of closeness of a representation or a measurement of a quantity to its actual value, whereas the precision of a measurement is the degree to which repeated measurements, under unchanged conditions, show the same results.

Distortion and noise introduced by the data converter will affect the accuracy of its outputs. Noise, assuming time-invariant distortion, may affect the precision of the conversion. A converter with n bits need not necessarily have the full accuracy or precision implied by n, not because the converter is not performing as desired, but because the desired performance does, in fact, *refer* to other parameters. So, in order to continue to talk about the resolution of the converter, in a broader sense, the concept of the *effective number of bits* (ENOB) has long been used as an alias for the *signal-to-noise-and-distortion ratio* (SNDR or, equivalently, SINAD). In the case of a pure,

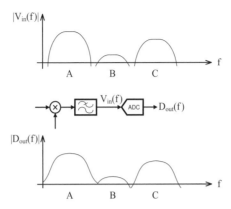

Figure 2.1 An example of a communication system in which, due to intermodulation distortion, spurious power corrupts adjacent channels.

full-scale digitized sinusoidal signal at a given frequency f_{sig} this is [52]

$$\text{ENOB}(f_{sig}) = \frac{\text{SNDR}_{dB}(f_{sig}) - 1.76 \text{ dB}}{6.02 \text{ dB}} \qquad (2.1)$$

where ENOB is a real number (almost always not an integer), expressed in bits, while SNDR$_{dB}$ is a dimensionless real number expressed in dB.[1] In order to be valuable, this expression also assumes a relatively high SNDR and a *quantization error* (sometimes also referred to as *quantization noise*) that can be reasonably approximated to a uniformly distributed error [52].

In Section 1.1, however, we have seen examples of a few applications and noticed how each one of these has different particular demands. Let us consider the example of the communication sub-system depicted in Fig. 2.1. In this figure an analog input v_{in} to an ADC consists of three adjacent communication channels, A, B, and C, of different power levels. Owing to (intermodulation) distortion caused by the ADC, undesired tones will be introduced into the digitized output D_{out}, affecting its accuracy and, in particular, causing more serious deterioration of the weakest channel B due to its relatively lower power level. In such an example, the system designer would want to specify the maximum acceptable deterioration of channel B. Therefore, given the power distribution of the channels over frequency, it will be meaningful to specify the corresponding *adjacent channel power ratio* (ACPR), namely the ratio between the (spurious) power introduced into the adjacent channel, due to distortion by the ADC, and the power of the main channel itself. Let us notice at this point that, although this specified ACPR could be made to correspond to a certain SNDR and, hence, using Eq. (2.1), to a corresponding minimum effective number of bits ENOB$_{min}$, this would not be very useful. That's not only because of all the assumptions that would need to be made about signals and their

---

[1] In this book, quantities expressed in dB will be explicitly marked as such; otherwise it will be assumed that their straight numerical value is being used.

power in order to go from one metric to another, but also because the opposite mapping from ENOB back to ACPR is not necessarily going to work. In fact, the ENOB alone (or the resolution in integer bits n) says nothing about the nature of the distortion mechanism and associated conditions, and therefore assigning a single specification to $ENOB_{min}$ would not insure that the desired ACPR specification is actually met. Clearly the ACPR *is* the desired specification that the ADC circuit designer needs to meet, characterize, and provide in the datasheet, while the ENOB will give only an indication regarding the minimum resolution of the ADC that can possibly address the application need and will, in fact, give a limited depth of discussion with the communication system designer interested in obtaining an accurate representation of the input signal.

Similarly, the sample rate f_s will also need to be properly used if considered as a performance metric. For example, if a signal with power between zero frequency and bandwidth BW is being converted, it will be important to pay attention to which converter architecture and anti-alias/reconstruction filter is being used in order to choose a proper f_s. If, for instance, a Nyquist-rate converter is used, then the absolute minimum f_s satisfying the sampling theorem will need to be $f_s = BW \cdot 2$. However, practical considerations for the design of the analog anti-alias/reconstruction filter will likely lead to choosing a much higher sample rate, for example $f_s = BW \cdot 5$ or higher. On the other hand, if an oversampled converter (e.g. a $\Delta\Sigma$ converter) is used, then, since the sample rate f_s, by design, is generally very high compared with the signal bandwidth BW, it will not be that unusual to have $f_s = BW \cdot 2OSR$ with $OSR = 8, 16, 32, \ldots$, where OSR is the *oversampling ratio*. In the end, then, the real specification is actually the signal bandwidth BW rather than the sample rate f_s, since the same signal could be equally converted by different architectures with possibly greatly different values of f_s. It is, in passing, interesting to remark that, contrary to a first guess, the most suitable choice of the converter–filter combination need not necessarily be the one with the smallest f_s with regard to cost, power consumption, complexity etc. Most current digital audio applications are striking examples of that since $\Delta\Sigma$ converters with large values of the OSR (and hence f_s) are universally used in this context, allowing also the design of practical, compact, and effective analog filters.

From the considerations and examples cited above it is clear, then, that the real ADC/DAC specifications one should consider are those characterizing the signals that need to be converted (such as the bandwidth BW and/or center frequency f_c to locate its power content) and those specifying their processing by the converters (e.g. ACPR, SNR, etc. to characterize accuracy etc.), rather than somewhat artificially concentrating on the sample rate f_s and resolution n alone.

2.2   How is performance being measured?

So what are the main performance metrics we should be concerned with? A brief summary of the most used ones is here informally provided only for convenience and to refresh the reader's memory. Very formal definitions for such metrics go beyond our aims and can easily be found in the literature.

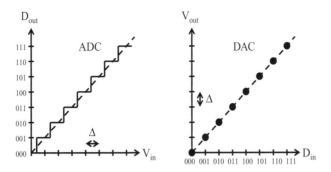

Figure 2.2 Ideal transfer characteristics of an ADC and a DAC.

First of all we should distinguish between *static* and *dynamic* metrics [12]. Static metrics refer to the context of signals at very low frequency, ideally at zero frequency (DC), while dynamic metrics refer to the case in which signals have high frequency.

2.2.1 Static metrics

In a static sense, the function of a linear data converter is to linearly map an analog V_{in} (or a digital D_{in}) input into a digital D_{out} (or an analog V_{out}) output. Therefore the ideal transfer characteristic consists of a staircase-shaped curve with all its identical steps aligned in a straight line as depicted in Fig. 2.2.

In the case of an ADC the steps lead to quantization errors because each digital output D_{out} represents a continuous range of values for its analog input V_{in}. In the case of a DAC, formally, each digital input D_{in} corresponds to a specific analog output V_{out}, and therefore one can say that no quantization error is introduced by the DAC; but that, of course, is true because D_{in} is already a quantized variable by definition.

Furthermore, if the characteristic spans positive and negative analog/digital values (as opposed to same-sign values), or if the analog or digital zeros fall on a defined point of the characteristic as opposed to falling on a transition point between a step and the next one, one can have various types of characteristics called *bipolar/unipolar* and/or *mid-rise/mid-thread*. Once again, we refer the interested reader to the open literature for a deeper discussion [54, 55].

In actuality the staircase and its steps can deviate from that in many different ways and the corresponding errors are quantified by

- offset error;
- gain error;
- differential nonlinearity (DNL);
- integral nonlinearity (INL).

The *offset error* quantifies the amount by which the actual characteristic is linearly shifted from its ideal position. The *gain error* quantifies the deviation of the slope of the

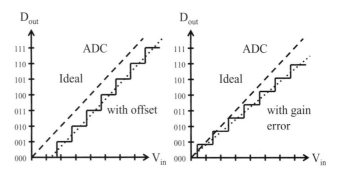

Figure 2.3 Offset and gain error in an ADC.

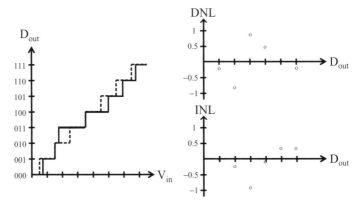

Figure 2.4 An example of INL and DNL for an ADC.

actual straight staircase from its intended slope. Offset and gain error are conceptually straightforward, as shown in Fig. 2.3 for the case of an ADC, although their formal definitions have some subtleties we will not explore here [12, 54, 55].

Both of these errors do not capture explicitly any deviation of the size of the steps from their intended value.

However, the size of the quantization steps can deviate from their ideal size Δ, also commonly referred to as the *least significant bit* (LSB) size. Individual deviations of each step from the ideal value Δ will lead to localized errors in the characteristic.

Such deviations are quantified by the DNL, which specifies how much each step differs from Δ. For example, referring to Fig. 2.4, let us consider the step width corresponding to $D_{out} = 001$. The actual step (solid line) is slightly smaller than the average LSB size. This small negative deviation corresponds to the small negative DNL point on the upper right-hand-side DNL plot. On the other hand, on observing the step width corresponding to $D_{out} = 011$, we see that this is considerably larger than the average LSB size, hence leading to a large positive point in the DNL plot for this code.

The minimum possible value for an ADC's DNL is clearly -1 since it corresponds to a step width that is -1 LSB smaller than the average step size, namely it corresponds to no step at all! This is the case of a *missing code*. On the other hand, a step size could be extremely large, hence having a DNL point larger than +1 is bad but possible.

For a DAC, on the other hand, since the output is analog we could have a non-monotonic output (i.e. the output will go up and down with increasing code instead of monotonically increasing as in the ideal case on the right-hand side of Fig. 2.2). Hence, since for DACs we measure the step sizes on the vertical axis, when there is a non-monotonicity problem then a DNL < -1 is actually possible.

Furthermore, again both for ADCs and for DACs, step-size errors also lead the transfer characteristic to move away from the ideal straight line. The net deviation of the actual characteristic from such an ideal straight line is quantified by the INL. As the nomenclature of these two errors suggests, the DNL at each step bears no memory of such deviation errors that have occurred at other steps and it is a localized error. On the other hand, the compounding effect of subsequent localized step-size errors, originating from one step to the next one and so on, progressively moves the characteristic away and back from its ideal straight line, hence leading to an integral error that is captured by the INL [56] (see also the ADC example in Fig. 2.4).

The gain error does not introduce nonlinearity. However, it can be an important error to account for in some applications such as, for example, in test and measurement applications and in some industrial applications. If necessary, the gain error is generally easy to compensate for by means of various circuit techniques and/or trimming or calibration.

The offset error, formally, does introduce nonlinearity since the superposition principle does not hold;[2] however, since it does not introduce harmonics and it too can be easily compensated for by means of various circuit techniques and/or trimming or calibration, it isn't a difficult matter to deal with.

On the other hand, static linearity errors affecting the shape of the transfer characteristic and quantified by the INL and DNL are actual nonlinearity. Such errors can be very important right at DC for some applications such as, for example, higher-resolution imaging applications (where they can lead to visible persistent undesired variations in brightness, color etc.). Furthermore, these introduce harmonics into the output and set a minimum of distortion that will degrade further, as frequency increases, since dynamic nonlinearity compounds it, making it progressively worse.[3] So, if we are trying to hit a high frequency distortion specification then we certainly need to insure that the static nonlinearity is at least adequate (equal to or lower than the maximum admissible distortion) to leave additional margin for the onset of further dynamic nonlinearity at higher frequency.

An actual noisy signal will wander around the transition points of the transfer characteristic and, hence, a sufficiently powerful noise, namely one that is comparable to the size of the LSB Δ, can "smear out" the DNL even to the point of hiding missing codes [55]. This smearing effect is actually exploited in *dithering* techniques to reduce

---

[2] This is related to the known difference, in mathematics, between strictly *linear functions* and *affine functions*.

[3] One could argue that, in general, the overall nonlinearity could diminish by fortuitous compensation effects at higher frequency leading to smaller distortion at high frequency than at low frequency. However, such events are rare and, from an engineering standpoint, we won't operate on the assumption that this will actually happen.

the harmonic distortion resulting from DNL errors [58, 59, 60]. Because of this masking effect due to noise, for converters whose input referred (thermal) noise is larger than Δ, the DNL is not a very useful metric [54].

The DNL and INL can be minimized by design. These errors often originate from a combination of mismatch of active and passive components as well as from circuit limitations such as finite gain of amplifiers in closed feedback circuits. Proper architectural choices (e.g. segmentation [61, 63]) together with layout techniques can help minimizing matching issues [33, 64], while circuit design techniques exist to cope with various circuit limitations (e.g. finite gain in amplifiers) [65] as well as matching limitations [66, 67, 68].

Trimming and calibration can be used to measure and compensate for the impairments leading to DNL/INL errors beyond what is otherwise achieved by circuit and layout design techniques [52, 69], as will be further discussed in Section 2.5.

2.2.2   Dynamic metrics

As the signal frequency increases frequency-dependent impairments of the converter's circuitry, which were negligible at lower frequency, become progressively more significant, further degrading the performance of the converter. Moreover, there exist performance metrics to characterize noise and distortion associated with specific applications. Some of these will be summarized here.

Converters' dynamic response can be characterized in the *time* domain and in the *frequency* domain. In the time domain square waves and rectangular pulses are the most common test signals (analog or digital) fed into the input of the converter. On feeding in a low-frequency pulse train or square wave, the shape of the output of the converter is observed in order to assess well-known classic step-response metrics such as *settling time*, *overshoot*, *slew rate* etc. as well as to spot slowly settling tails etc. [52, 33]. In imaging/graphic applications this type of test corresponds to graphic pattern tests (e.g. checkerboard tests showing possible smearing or contrast problems at color transition borders) whereby the quality of the output image is directly observed [70, 71]. In test and measurement applications, pulse response problems can lead to baseline or gain wander [73].

Another useful time-domain test is the *beat frequency* test, whereby a square wave at a frequency f_{in} that is very close to the sampling rate f_s is fed into the input of the converter. Because of aliasing this results in a low-frequency ($f_{out} = f_s - f_{in}$ – the "beat frequency") output periodic response, which is then rich in details observable in the time domain before the waveform repeats itself. However, since the converter has been exercised at very high frequency, evaluating the shape of this output can be useful as a means to uncover some dynamic limitations not discovered with the previously discussed low-frequency square-wave input. That includes timing errors (which show up as distortion on the beat pulse), missing codes, etc. [53, 73].

In the frequency domain, typical test signals are single sine waves (a tone), pairs of sine waves (two tones), multiple tones, and narrowband signals with roughly constant power within the band itself.

The metrics discussed in the following vary in value depending on the amplitude and on the frequency of the signals involved [74]. So quoting a certain metric requires describing the conditions under which it is provided. In the case of ADCs, when these conditions aren't explicitly stated, it is often implied that the input signal is at full scale amplitude and at the Nyquist frequency ($f_s/2$).

In the case of DACs, especially the current-steering DACs discussed in Chapter 4, on the other hand, most dynamic metrics vary dramatically in value depending upon the output signal power (P_{sig}), frequency (f_{out}), and output rate (f_s), so specifying the conditions under which a metric is quoted is absolutely mandatory.

The most common metrics are certainly the *signal-to-noise ratio* (SNR) and the *total harmonic distortion* (THD), namely the power ratios

$$\text{SNR} = \frac{P_{sig}}{P_{noise}} \tag{2.2}$$

and

$$\text{THD} = \frac{P_h}{P_{sig}} \tag{2.3}$$

where a pure sine-wave signal of power P_{sig} is assumed, P_{noise} is the power of the noise, and P_h is the power of the harmonics.

Other metrics are associated with these, such as the already-cited *signal-to-noise-and-distortion ratio* (SNDR) (also known as the SINAD)

$$\text{SNDR} = \frac{P_{sig}}{P_{noise} + P_h} \tag{2.4}$$

and its reciprocal $\text{THD} + N = 1/\text{SNDR}$, which is more commonly encountered in digital audio applications [75].

Specific distortion metrics include the individual nth-order *harmonic distortions* (HDs)

$$\text{HD}_n = \frac{P_n}{P_{sig}} \tag{2.5}$$

where n specifies the order of the harmonic (second harmonic, third harmonic, etc.) and P_n is the power of the corresponding harmonic. Also, the *spurious-free dynamic range* (SFDR) is often used to identify the largest spurious tonal signal since it is the ratio

$$\text{SFDR} = \frac{P_{sig}}{P_{hmax}} \tag{2.6}$$

where P_{hmax} is the power of the largest undesired signal (frequency-related or unrelated to the useful signal). Related to the SNR is also the *dynamic range* (DR), which is the power ratio between the largest and smallest detectable signals (the smallest being essentially the noise, which can either be the quantization noise or the actual device noise, depending on the specific case).

Let us consider an example to fix the ideas. Figure 2.5 shows the spectrum of the output of an ADC sampling at 100 MS/s, an input signal at $f_{sig} = 29.35\,\text{MHz}$ and second and third harmonics aliased to $f_2 = 41.31\,\text{MHz}$ and $f_3 = 11.96\,\text{MHz}$,

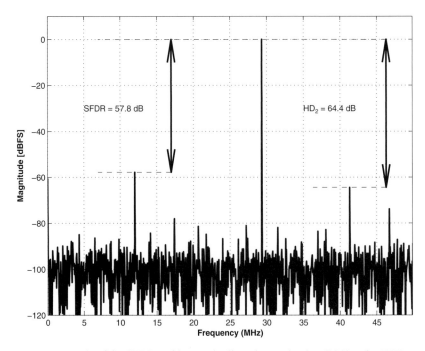

Figure 2.5 An example of the SFDR and harmonic-distortion evaluation (2048-point FFT).

respectively. Their normalized powers are $P_2 = -64.40\,\text{dB}$ and $P_3 = -57.80\,\text{dB}$, respectively. Therefore, since the power of the fundamental is $0\,\text{dB}$, then, working with logarithmic units, we have $\text{HD}_{2\text{dB}} = P_1 - P_2 = 0 - (-64.40) = 64.40\,\text{dB}$ and $\text{HD}_{3\text{dB}} = P_1 - P_3 = 0 - (-57.80) = 57.80\,\text{dB}$. Furthermore, since in this example the third harmonic is the strongest spurious signal, the SFDR coincides with the third-harmonic distortion: $\text{SFDR}_{\text{dB}} = \text{HD}_{3\text{dB}} = 57.80\,\text{dB}$.

A very important noise performance metric is the *noise spectral density* (NSD) [76]. This is a measurement, in the frequency domain, of the noise per unit of bandwidth (spot noise) at a specified frequency. This can be specified in direct units, such as $\text{nV}/\sqrt{\text{Hz}}$ [5, 33] or in logarithmic units, namely in dB/Hz, and typically it is specified in the presence of a $0\,\text{dB}$ full-scale signal at a specified frequency [77].

Given the observation bandwidth BW and the NSD it is then possible to relate these to the SNR within this bandwidth by noise power integration. Assuming a flat NSD over the BW, this is straightforward using logarithmic units:

$$\text{NSD}_{\text{dB/Hz}} = -(\text{SNR}_{\text{dB}} + 10 \cdot \log(\text{BW})) \qquad (2.7)$$

Related to the ENOB is the *error resolution bandwidth* (ERBW). The ERBW is the input signal frequency at which the output ENOB of a converter drops by 0.5 bits. It should be noticed that the ERBW is often larger than the Nyquist frequency $f_s/2$ in many ADCs.

Single-tone tests can be useful when dealing with wide bands and/or in the context of Nyquist-rate converters. In these two cases all the spurious signals fall within the frequency range of interest. For example, harmonics that may fall beyond the Nyquist frequency $f_s/2$ are aliased back within this range and therefore are accounted for. However, when dealing with narrow bands, or when using oversampled converters, even lower-order harmonics such as the second- or the third-order harmonic could have sufficiently high frequency to fall outside the observed frequency range (or not to be aliased back within the observed frequency range). In the event that these harmonics fall outside the observation range, clearly, one could be led to the erroneous conclusion that the system under observation achieves low distortion.

To address this issue two-tone tests are performed. Namely, two sine waves, usually of equal power and similar frequencies f_1 and f_2 and called *fundamental frequencies* are fed in as an input. Distortion (in this case often referred to as *intermodulation distortion* or *IMD*) will then introduce a number of additional tones at frequencies $k_1 \cdot f_1 + k_2 \cdot f_2$, with k_1 and k_2 arbitrary integer values. These tones are called *intermodulation products* of order $n = |k_1| + |k_2|$. Only odd-order products are close to the fundamentals; for example, for $f_1 = 10\,\text{MHz}$ and $f_2 = 11\,\text{MHz}$ two third-order products are at $2f_1 - f_2 = 9\,\text{MHz}$ and $2f_2 - f_1 = 12\,\text{MHz}$ (while the other two third-order products are distant: $2f_1 + f_2 = 31\,\text{MHz}$ and $2f_2 + f_1 = 32\,\text{MHz}$) while the second-order products $f_1 + f_2 = 21\,\text{MHz}$ and $f_2 - f_1 = 1\,\text{MHz}$ are distant [78].

This can actually be generalized to an arbitrary number of fundamental tones $f_1, f_2, f_3, \ldots, f_N$, which, due to IMD, will lead to a family of products $k_1 f_1 + k_2 f_2 + \cdots + k_N f_N$. Multi-tone tests are indeed also performed.

The usefulness of this type of test, as previously hinted, comes from the fact that some of the products (e.g. two out of four of the third-order products in the case of the two-tone test) will fall very close to the test tones and hence will provide the desired information on the distortion. The ratio of the fundamentals to the products defines the *intermodulation distortion* (IM) of order n:

$$\text{IM}_n = \frac{P_{\text{sig}}}{P_{\text{IP}_n}} \tag{2.8}$$

where P_{sig} is the power of one of the two identical fundamentals while P_{IP_n} is the power of one of the products. So, for example, the third-order intermodulation distortion IM_3 can be obtained as the ratio of the power of the fundamental at f_1 and the power of the product at $2f_1 - f_2$.

If the nonlinearity of the converter can be reasonably approximated by a power series then, on increasing the signal power of the fundamentals, the signal power of the products will increase exponentially with the order of intermodulation. So, for example, if the signal power of the fundamental is increased by 10 dB, the signal power of the third-order intermodulation products will increase by 30 dB. This model works reasonably well for signals that are relatively close to full scale. However, as signals drop in power and quantization becomes increasingly dominant over "weak" continuous nonlinearities that can be modeled with power series (such as polynomials or Volterra series), distortion products do not scale following the previously described law. Unfortunately, the power of these products drops at a slower pace, as it will be described a bit more later.

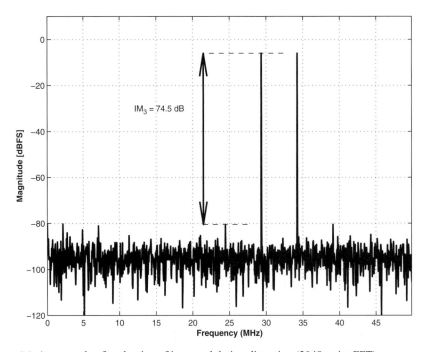

Figure 2.6 An example of evaluation of intermodulation distortion (2048-point FFT).

Let us consider an example. Figure 2.6 shows the spectrum of the output of an ADC with a two-tone input and a sample rate of 100 MS/s. The frequencies of the two tones are $f_1 = 29.35$ MHz and $f_2 = 34.23$ MHz and both have a power $P_{\text{sigdB}} = P_1 = P_2 = -6$ dB. The third-order intermodulation products are clearly visible at $2f_1 - f_2 = 24.47$ MHz and $2f_2 - f_1 = 39.11$ MHz, and each of them has a power $P_{\text{IP}_3\text{dB}} = -80.5$ dB. The third-order intermodulation distortion is easily computed using logarithmic units as $\text{IM}_{3\text{dB}} = P_{\text{sigdB}} - P_{\text{IP}_3\text{dB}} = -6 - (-80.5) = 74.5$ dB.

An important metric often used in conjunction with intermodulation is the *intermodulation intercept point* (IP) of order n. Let us imagine, for instance, feeding two tones (the "fundamentals") with combined amplitude close to full scale into the input of our converter and measuring the corresponding IM_3. Let us increase the power of the two tones, without reaching full scale (in order to avoid clipping), and record the corresponding increase in the third-order products at the output. As stated earlier, in the range where the distortion of the converter can be represented by using a power series (namely, for a range of the inputs just below full scale, where signals are small enough that hard-limiting clipping has not yet occurred and where they are large enough that quantization is not the primary source of distortion) it is possible to extrapolate and draw straight lines on a log–log scale representing the power of the fundamentals and the corresponding power of the third-order products at the output of the ADC for increasing input power levels as shown in Fig. 2.7.

The two extrapolated power curves intercept at a point called the third-order intermodulation output intercept point, OIP_3. Correspondingly, the horizontal coordinate for the intersection is the input-referred intermodulation intercept point, IIP_3.

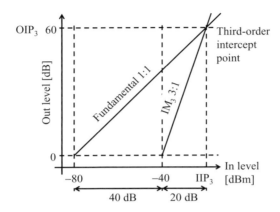

Figure 2.7 Definition of the third-order intermodulation intercept point.

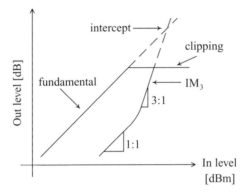

Figure 2.8 Actual power trends for fundamentals and IM_3 in an ADC.

It should be again emphasized that the intercept point is a mathematical abstraction. It, in fact, has a number of limitations that it is now important to highlight.

If what has just been described is applied to a standard linear analog circuit such as an amplifier then, as the input amplitude approaches full scale or the supply voltage, the output begins soft-limiting/compressing. Hence, for example, the power curve of the fundamental does not keep the unity slope but starts to flatten. The two curves, in fact, simply do not meet for realistic power values. But, again, the extrapolation that leads to the intercept point is meaningful since the power trends, as sketched, allow one to predict the distortion over a large span of input levels.[4]

As has previously been pointed out, in a data converter, the concept of an intercept point needs to be taken with a grain of salt. Let us refer to the example of power trends shown in Fig. 2.8. As stated above, when the input amplitude reaches full scale the output is clipped (in fact, in some converters, it may create undesired overload conditions that can lead to lengthy erroneous output sequences), hence the distortion dramatically and suddenly increases. Furthermore, if the input amplitude is appreciably lower than full scale then quantization effects can even make the power of the distortion tones change

---

[4] Let us also remark that, since the distortion is frequency-dependent, these curves will change with frequency.

at the same rate as that of the fundamentals (i.e. for a 1 dB drop in fundamental power one will see a 1 dB drop in third-order intermodulation tones instead of a 3 dB drop as previously mentioned). So the trends sketched in Fig. 2.7 are not actually representative of what is observed with an ADC (or a DAC) when considering a large range of inputs. For example, when an IIP_3 is quoted for a converter, it will be important to specify the entire set of conditions over which it is discussed (i.e. the input power and, of course, the frequency).

Although, as explained above, it is not very meaningful to use intercept points directly with data converters, these are quite common metrics in the context of communication systems (where converters are increasingly crucial components) and it is important to understand their meaning and the limitations associated with them.

It is also important to mention that, although the concept of an intercept point has here been recalled in the context of two-tone tests and intermodulation distortion, it is also possible to analogously define another intercept point for single-tone tests [78] by considering the intersection between the power of the fundamental and that of a chosen harmonic for increasing input signal power. Once again, as in the case of intermodulation, this intersection is theoretical, being obtained by the linear extrapolation of these powers when they are rather limited and the circuit is still behaving as a weakly nonlinear circuit.

Going from two-tone tests to multi-tone tests [12] and, more importantly, to narrow-band signal tests, one more important performance metric is recalled here. That is the already-cited *adjacent channel power ratio* (ACPR) or *adjacent channel leakage ratio* (ACLR). In this type of test, a narrowband signal, with center frequency f_A and bandwidth BW_A and with uniform power within BW_A is fed into the input of the circuit under test. The signal at the output will contain the response A to this "channel" as well as some spurious power falling into the adjacent frequency ranges (the adjacent "channels"). Just like in all other spurious signal ratios, the ratio between the intended signal (A) power P_A and the spurious power, in this case the total spurious power measured in the adjacent channel frequency range P_{AB}, constitutes the ACPR, expressed in dBc:

$$\text{ACPR} = \frac{P_A}{P_{AB}} \tag{2.9}$$

The bandwidth/channel spacing and the power P_A (as well as the specific actual narrowband test signal) vary depending on the application.[5] ACPR tests can be done with a single active channel (as just discussed) or with multiple active channels as previously mentioned in the communication application example in Section 2.1.

Also, adjacent channel "leakage" can occur due to distortion as described in the previous discussion on intermodulation, but, particularly at very high frequency, it can also result from aperture sampling error when a jittery sampling clock effectively phase-modulates the input signal. This is often characterized by "noise skirts" at the edges of the test channels with a shape resembling the sampling clock's phase noise contour [77, 79].

---

[5] This type of test is primarily used in communication applications, so different values are used for different standards like WCDMA, CDMA2000, TDSCDMA etc.

2.3   Figures of merit

The metrics summarized in Section 2.2 are only some of the most commonly used among a considerably larger universe of performance parameters. These metrics aim to characterize actual physical parameters, or their ratios, in order to specify, design, and verify converters for targeted applications.

On the other hand, provided that a converter meets the desired specifications on the basis of the above-cited performance metrics, it is worthwhile wondering how well that one specific design is optimal with respect to some other engineering parameter, which may be power efficiency, available signal bandwidth etc.

Figures of merit (FOMs) are then introduced with this aim, allowing (or at least attempting to allow) also comparisons between various converters which could differ widely in architecture, application, and specifications.

Focusing on ADCs first, one of the most commonly used FOMs is that called "Walden's" FOM [80]:

$$\text{FOM}_\text{W} = \frac{P}{f_\text{s} \cdot 2^{\text{ENOB}}} \tag{2.10}$$

where P is the power dissipation of the ADC. FOM_W is an efficiency parameter that aims to provide a measure of the energy required by the ADC to perform a conversion and it is expressed in *picojoules per step*. FOM_W could be justified empirically after surveying a large population of ADCs, although, theoretically, it suffers from an important limitation. Namely, it suggests that in order to double accuracy (i.e. to add one more bit to the ENOB) the power consumption P will only need to double. However, if, for example, the ADC's ENOB is limited by thermal noise (i.e. $k_\text{B}T/C$ limited) only,[6] then, in order to decrease the noise floor by 6 dB (so that the ENOB increases by 1 bit), C needs to quadruple. If f_s is kept constant, then the ADC's dynamic power consumption will need to increase by a factor of four, not a factor of two, as implied by Eq. (2.10).

Minor variants of FOM_W include (a) replacing f_s with $2 \cdot \text{BW}$ in order to account for oversampled converters [81, 82, 83]; (b) replacing f_s with the ERBW to remove the Nyquist-rate emphasis altogether and instead focusing on the frequency performance of the converter itself [29]; (c) replacing 2^{ENOB} with the absolute value (i.e. not in dB) of the SNDR or SNR [82, 83]; (d) considering the reciprocal of Eq. (2.10) as a FOM since FOM_W as defined in Eq. (2.10) gets smaller for more efficient ADCs [29].

To address the theoretical limitation cited above, for thermal-noise-limited ADCs (shot-noise-limited ADCs show the same type of trend), modified FOMs with a squared accuracy term have been proposed by various authors [29, 83]. One is the following:

$$\text{FOM}_2 = \frac{P}{2\text{BW} \cdot \text{SNTR}^2} \tag{2.11}$$

where the accuracy term SNTR, the *signal-to-thermal-noise ratio*, now appears raised to the power of 2. Another variant, increasing in value for better-performing designs,

---

[6] With negligible distortion and quantization noise.

expressed in dB and using the SNDR (in dB) as the accuracy term, is [81, 69]

$$FOM_3 = SNDR_{dB} + 10 \cdot \log \left(\frac{BW}{P} \right) \tag{2.12}$$

Surveys of ADCs' performance have been published over the years [28, 80]. In fact, Professor Murmann of Stanford University has been regularly updating a collection of such data points on the basis of results published at key scientific conferences [85]. Inferences on ADC performance trends can be done analyzing such data also with the aid of the above-reported FOMs [83, 85]. This is, in fact, insightful and such a discussion is the topic of Chapter 5.

Much less attention, in general, has been paid to benchmarking DACs with FOMs. Focusing on high-speed (e.g. $f_s > 100$ MS/s) current-steering DACs only, many different FOMs are found in the literature, and a few of them will be briefly reported here. Some of the "randomness" in DACs' FOMs originates from the fundamental question of what performance metric(s) can more broadly capture the value and the technical challenges of a given design when considering such a very diverse and wide application space. Furthermore, some of the performance parameters tend to vary much more, over frequency, than in ADCs. For example, the linearity of the output of this type of DACs degrades very rapidly with the output frequency. Therefore, particularly in wideband applications, it is important to capture how well the DAC can meet the desired linearity at a certain frequency. Moreover, the linearity is also strongly dependent on the output swing, and the output swing is important from a noise standpoint. It can then be seen how the complexity of capturing all these important factors and requirements easily creates a conflict with the desire to define a simple FOM that can give a unique score to such a multi-dimensional problem.

The first FOM we will consider resembles, in essence, those previously discussed in the context of ADCs. This FOM [86, 87] has the physical dimensions of the reciprocal of energy:

$$FOM_{vdb} = \frac{2^n \cdot BW_N}{P} \tag{2.13}$$

where P is the total power dissipation, n is the number of bits, and BW_N is the frequency at which the SFDR drops by 6 dB from its DC value. This FOM does not capture the output swing or the amount of power delivered to the output load. Because of the former issue, this favors DACs with low output swing, which would provide a more linear output but won't necessarily supply the load with the desired signal power.

A better FOM is then the following one [88]:

$$FOM_G = \frac{V_{swing} \cdot BW}{P} \cdot 10^{SFDR/20} \tag{2.14}$$

where V_{swing} is the output voltage swing, while BW is the frequency of the tone used to measure the SFDR. However, as mentioned above, the output distortion is strongly frequency-dependent and hence so is the SFDR. Furthermore, FOM_G has the physical dimensions of the reciprocal of a charge, which doesn't offer a clear sense of what the FOM is actually aiming to assess. A further improved FOM is then the following one [89]:

$$\text{FOM}_C = \frac{2^{(\text{SFDR}_{\text{DC}}-1.76)/6.02} \cdot 2^{(\text{SFDR}_{0.5 \cdot f_s}-1.76)/6.02} \cdot f_s}{P - \frac{1}{2} \cdot I_{\text{Opk}}^2 \cdot R_L \cdot D} \tag{2.15}$$

where SFDR_{DC} is the SFDR measured for an output tone at zero frequency, $\text{SFDR}_{0.5 \cdot f_s}$ is the SFDR measured for an output tone at the Nyquist frequency, I_{Opk} is the peak output current, R_L is the output load resistance, and D is the duty cycle of the output (hence $D = 1$ for non-return-to-zero DACs, while $D = 0.5$ for return-to-zero DACs with half-clock duty cycle). It can be seen that also this FOM somewhat resembles the previously discussed ADC FOMs: the first two terms in the numerator provide an assessment of the linearity of the DAC by means of its SFDR at DC and at the Nyquist frequency, while the denominator assesses the power internally dissipated by the DAC by subtracting the power delivered to the load from the total power of the overall "DAC plus load" system. In this FOM the role of the output swing is indirectly captured through I_{Opk}. Emphasis is instead placed on the transfer of signal power to the load R_L. Power transfer is the primary emphasis of the last DAC FOM presented in this section, which is indeed a normalized power efficiency [90]:

$$\text{FOM}_L = \frac{P_{\text{pk}}(R_L)}{0.25 P_{\text{supply}}} \tag{2.16}$$

where $P_{\text{pk}}(R_L)$ is the available power for the load while the factor 0.25 in front of the power provided by the supply P_{supply} is used in order to allow the theoretical maximum of this ratio to be 100%.

Before concluding this discussion we should briefly mention, without digressing too much, something that concerns the computation of FOMs. Specifically, when comparing the FOM values reported in papers and technical reports the reader should be warned about the consistency (or lack thereof) in the computation of such values. In some papers the power consumption P used in the computation may include all circuit blocks. In other cases, for example, the power consumed by necessary ancillary blocks such as an on-chip reference voltage buffer, or the digital output drivers delivering the output bits to other devices on the same PCB, or a large digital block performing calibration functions etc., may be left out and, therefore, artificially give a more favorable FOM. Moreover, certain architectural choices may also accidentally provide misleading FOM values. For example, a charge-redistribution architecture (e.g. those used in some successive-approximation converters) with a large capacitance would provide a low $k_B T/C$ noise and good matching. However, depending on the specific converter architecture, it will also present a hefty load to the input driving source and to the voltage reference driver. The corresponding power involved need not necessarily be accounted for in the FOM computation, but it would be a "necessary burden" shifted from the ADC to the rest of the system. Furthermore, in one paper the SNDR may be computed for an input sine close to the Nyquist frequency, while in another paper it may be computed under other, more favorable, conditions.

In conclusion, the key message of this brief digression is that, regardless of the endless debates on how suitable (or unsuitable) a certain FOM may be to assess the value of a

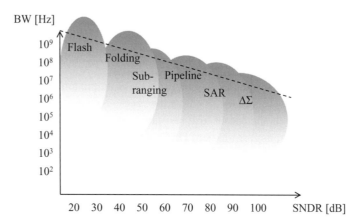

Figure 2.9 Traditional ADC architectures in the BW versus SNDR plane.

certain design because it takes into account or neglects important performance metrics, the reader should also be warned about the variability in the computation methodology used to determine the value of the chosen FOM. FOMs are certainly valuable and, indeed, very useful data points. However, anyone who uses them to make assessments and comparisons should exercise care and technical maturity before using them to draw conclusions.

2.4   Classic architectures

Once the specifications that the converter application requires have been identified and formalized as discussed in Section 2.1 the choice of converter architecture needs to be made.

This is actually a very complex problem both because there are several engineering variables involved in this choice (e.g. the process technology, the power consumption, the silicon area, the conditioning circuitry between the converter and the remainder of the system etc.) as well as because often multiple architectures may be suitable to meet the target requirements, not to mention that organizational factors may play a strong role in this choice. For example, the experience of the design team with a particular architecture or the availability of an existing IP that could be re-used to shorten the development time would often take a higher priority than, for example, choosing the architecture with the most optimal power or area for the case at hand.

Once that has been said, it is possible to roughly sketch a "map" of where classic converter architectures fit with respect to the main key performance metrics. Figure 2.9 shows such an empirical chart for classic ADCs in the signal bandwidth[7] BW versus signal-to-noise-and-distortion SNDR space at the time of writing.

---

[7] An implicit assumption here is that the signal lies between DC and BW. Bandpass ADCs, for example, are not considered here.

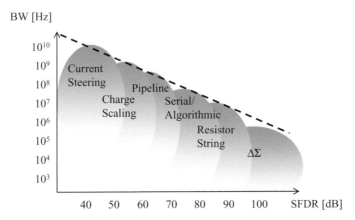

Figure 2.10 Traditional DAC architectures in the BW–SFDR plane.

Given the vast range of the two parameters, each architecture is shown to occupy a limited region (sketched as a fading ellipse) of such a space. Consistently with what has just been pointed out, since multiple architectures can be suitable for very similar specifications, it can also be seen how many regions show overlap between multiple architectures.

A diagonal dashed line cutting through the BW–SNDR space outlines a practical border line on the present state of the art in terms of ADC performance. The region below it represents what it is at present feasible to design, while hitting specifications above the dashed line is increasingly non-trivial or simply not yet feasible. Interestingly, if we consider the well-known *sampling aperture jitter formula* [52, 81, 91] for the SNR solely due to sampling an input sine wave of amplitude A at frequency f_{in} with a clock with rms jitter σ_j,

$$\text{SNR}_{dB} = 20 \log \left(\frac{1}{\sqrt{2} A \pi f_{in} \sigma_j} \right) \qquad (2.17)$$

and we plot its contour lines (assuming that the normalized full-scale amplitude is $A = 1\,\text{V}$) in the chart of Fig. 2.9 for decreasing values of σ_j, these correspond to diagonal lines, parallel to the dashed line, gradually going further and further away from the origin. The dashed line in Fig. 2.9 corresponds to $\sigma_j \sim 1\,\text{ps}$ (rms).

This clearly indicates one of the fundamental challenges in pushing the speed–resolution performance and transcends the ADC architecture (i.e. the actual conversion process) insofar as it relates only to sampling.

A similar chart can be made for DACs as well. Here, as pointed out in Section 2.3, it is hard to choose which metrics can be more encompassing for such a broad space of cases. A somewhat arbitrary choice has then been made in the chart shown in Fig. 2.10, where, similarly to what was done for the ADCs, many classic DAC architectures have been mapped into the BW–SFDR plane. It should also be remarked that the extent of the overlap between some of these architectures is, in actuality, greater than what is depicted.

Even though distortion (characterized by the SFDR in Fig. 2.10) is here considered instead of the SNDR, a similar fundamental trade-off between signal frequency and dynamic range of the same nature as Eq. (2.17) exists for DACs as well. For example, in some cases, even if extremely low distortion could be achieved in principle, a noise specification would also exist in conjunction with a given design. Besides, the SFDR would ultimately be limited by the noise floor once all harmonics are buried in the noise. Some very-high-performance $\Delta\Sigma$ DACs used in digital audio applications approach the latter case. Therefore, in general, it is not surprising to find a similar diagonal border line in this chart as well.

Another important factor that plays an important role in the speed–resolution trade-off of both ADCs and DACs has to do with fundamental architectural compromises, in particular, the general design compromise between operating speed and complexity (and accuracy, unfortunately) in circuit topologies. On comparing Fig. 2.9 and Fig. 2.10 some common elements emerge. In both cases, as we move along the dashed lines from right to left toward higher and higher signal frequency, each converter architecture increasingly becomes constituted of multiple, topologically simpler, repeated blocks. Moreover, open-loop circuits take over from closed-loop circuits and systems. Also, oversampling decreases and gives way to Nyquist-rate converters.

For instance, in the case of ADCs, a $\Delta\Sigma$ modulator provides a high level of resolution by exploiting both noise-shaping and oversampling [81]. Noise-shaping is based on feedback and allows trading off of speed for dynamic range, while oversampling adds to the effect of noise-shaping by exploiting time-averaging. Because of the relatively lower frequency of the processed signals the relative complexity of some of the circuit blocks can be increased to better fit the need to handle the more linear signals, especially in the sections that are closer to the input and the summing node. As we move from the $\Delta\Sigma$ ADCs to the SAR ADCs, feedback (sensing–mixing) is still an important element in the operation of the converter. Each of the n iterations necessary to complete the conversion cycle requires the determination of each bit value (sensing) and, subsequently, the internal synthesis of the next approximation to the input signal, allowing the following finer comparison.

As we move from an SAR ADC to a pipelined ADC, at a high level, the principle of the conversion process has not changed that much; however, the iterative process that was cyclically performed by a single DAC-comparator system in the SAR is now "unfolded" into a cascade of synchronized subsequent sub-DAC/sub-ADC stages in the pipelined ADC, each resolving one or more bits and operating in a pipelined fashion over the input sampled time-series. Once again, on going from the SAR to the pipeline, the degree of feedback in the system has diminished, and the individual complexity of some of the building blocks has also diminished (for example the multiplicative DAC of the pipeline need only synthesize a signal of a few bits, whereas the DAC in the SAR ADC needs to account for all of the desired bits at once), hence allowing the design of higher-frequency circuits. Similar considerations can be made as we move toward various other algorithmic ADCs such as the sub-ranging ADCs, which operate under the principle of first identifying the coarse range within which the input signal lies (again, sensing) followed by a finer quantization within that range. Once again, there is less feedback, lower complexity of individual building blocks (however, there is a higher tally of them), and higher speed, but also a lower dynamic range.

In folding-interpolation ADCs the coarse quantization and the fine quantization happen independently and in parallel. Hence the architecture has given up yet another degree of feedback to gain in speed of operation. Also, the topological complexity of the individual blocks is also decreased further in that folding and pre-amp stages made of combinations of elementary differential pairs have considerably lower complexity than, for example, the amplifiers used in the multiplicative DACs of pipelined converters or $\Delta\Sigma$ ADCs. At the same time, the number of such elementary building blocks (folding circuitry, local comparators etc.) is again increased, becoming, for example, far greater than the number of (possibly multi-bit) cascaded stages of a pipelined ADC. In a folding–interpolating ADC most feedback is localized to only some of the individual blocks in order to limit parameter spread or locally improve linearity.

Lastly, flash ADCs represent the most open-loop architecture, and are conceptually the simplest of all architectures. That allows this architecture to be, without doubt, the fastest of all. On the other hand, giving up on feedback and its ability to control parameter spread and linearity means that the accuracy of these converters becomes rather limited. Furthermore, as the multiplicity of the basic building blocks (and the power and area required by them) increases exponentially with the resolution n ($2^n - 1$ comparators are required), practical flash ADCs are seldom designed for more than 8 bits (and hardly sustain a commensurate ENOB at high input signal frequency).

Very similar considerations can be made for DACs and so, for the sake of conciseness, we will simply compare current steering DACs with $\Delta\Sigma$ DACs. In fact, current-steering DACs have many conceptual similarities to flash or folding–interpolating ADCs in that the architecture of current-steering DACs is composed of multiple open-loop differential pairs operating all in parallel to synthesize the output. This large degree of parallelism, combined with a very-high-speed topological design style focusing on creating the smallest possible circuits with the minimum associated parasitics allows this type of DACs to reach very high speeds of operation. At the same time, similarly to what happens in flash and folding ADCs, also in current-steering DACs small device sizes and lack of feedback conspire against device matching, resulting in limited DC linearity. At high frequency, even small parasitics affect the linearity and the intrinsically low complexity does not allow for many effective ways to compensate for the impairments and continue to sustain the performance.

In $\Delta\Sigma$ DACs, analogously to their ADC counterparts, feedback (noise-shaping) and oversampling combine to provide high linearity and dynamic range but for rather small output frequencies, although in most $\Delta\Sigma$ DACs (even when using multi-bit architectures) the analog part of the converter is rather simple and somewhat reminiscent of some of the building blocks used in current-steering DACs. The multiplicity of such analog blocks is extremely small, in contrast to current-steering DACs.

So, once again, architectural trade-offs in ADCs and DACs play a significant role in determining the speed–accuracy trade-off visible in the charts of Fig. 2.9 and Fig. 2.10.

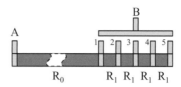

Figure 2.11 An example of trimming a polysilicon resistor.

2.5   Trimming and calibration

Various device and process technology impairments limit the performance of the converters. Some of these are addressed by choosing a suitable converter architecture, like those briefly recalled in the previous sections, and by properly specifying/designing the internal interconnections and the function and performance of the constituent blocks. Others are dealt with within the actual building blocks' implementation using classic circuit and layout design techniques [33, 64], allowing one to improve impedance ratios, frequency responses, noise, device matching etc. However, other alternatives are also available to deal with the effects of parameter spread, limits in components' precision, finite gains etc. These include *trimming* and *calibration* techniques.

In order to introduce the general idea, let us consider a simple example. Let us say we need a resistor with an absolute resistance of $R = 100 \pm 5\,\Omega$. Unfortunately, a regular manufactured polysilicon resistor often has a precision of the order of about 10%, which means that, if we simply manufacture a nominal resistor of 100 Ω, then, due to parameter spread, we could obtain anything between 90 Ω and 110 Ω. To control this value to lie within a narrower range we could manufacture the resistor as shown in Fig. 2.11, namely as a series of a large resistance R_0 and smaller resistors of resistance R_1. The desired R is obtained between contact A and contact B by shorting only one of the five openings (1, 2, 3, 4, and 5) between the contacts on the right-hand side. If R_0 is set to a nominal value of $R_0 = 90\,\Omega$ and the smaller resistors are set to $R_1 = 5\,\Omega$ then, neglecting the contact resistances, the nominal value of R could be set as $R = 90, 95, 100, 105$, or $110\,\Omega$, depending on which contact is shorted. However, now the precision has improved to the desired level. For example, if the manufacturing run leads to a negative skew of the resistivity by +10%, then $R_0 = 99\,\Omega$, and we can obtain our desired R by shorting contact 1 ($R = R_0 = 99\,\Omega$, while $R_1 = 5.5\,\Omega$). If, on the other hand, the manufacturing leads to a positive skew of, say, −8%, then $R_0 = 82.8\,\Omega$, and we can obtain our desired R by shorting contact 4 ($R_1 = 4.6\,\Omega$, then $R = R_0 + 3R_1 = 96.6\,\Omega$).

In summary, in order to obtain the desired precision we measured the desired parameter (e.g. by directly measuring the resistance of R_0 or the resistance of another similar resistor on the same die or another parameter indirectly related to it), we then took corrective action (i.e. we decided which one of the five contacts would have led to the desired value for R), and, finally, in order to be able to use the resistor at its desired value, we held the effect of the corrective action (by permanently shorting one of the five contacts).

So, once again, at a high level, trimming and calibration techniques entail three steps:

(1) measuring one or more parameters of the device that may need to be altered;

(2) corrective action on the device that changes the parameters to the desired value (possibly verifying that the parameter has been changed to the desired value);

(3) holding the corrective action (with the aim of holding the outcome of the parameter correction), for example, storing a record of it in some form of memory residing on the device or in a place directly accessible by the device during operation.

Let us now broaden this discussion a bit. First of all, although the two words can be synonymous in colloquial use, we will make a distinction below [92].

Trimming is performed during device production, at the factory, before it is shipped to the final user. Therefore, once trimming has been performed by the manufacturer, it cannot be changed and subsequent drifts in the trimmed parameters will not be corrected. So, if, for example, due to temperature changes, or aging, or device stress etc., the parameter varies, the device cannot be re-trimmed to counteract it. This needs to be pro-actively accounted for when the trimming strategy is devised and the final trimmed value is decided. Referring to the prior example of the poly resistor, this means that the values of R_0 and R_1 and the number of trimming resistors R_1 will need to be decided very carefully upfront. Also, the trimming algorithm used to measure and correct the parameter (the resistance value in this simple example), and the trimming environment (e.g. the parameter could be trimmed at wafer level, before the dies are diced; or it could be trimmed at die level, before the die is packaged; or it could be trimmed after packaging; in some rare cases, the manufacturer may agree to provide untrimmed devices to the final user, who will directly perform the one-time trimming at the destination), will need to be carefully thought through.

Various types of trimming are used nowadays and vary according to the method used to correct and set the final value of the desired parameter. Referring again to the example of the poly resistor, all five contacts could be normally closed before trimming and then, after trimming, all contacts that need to be open may be blown up like electrical fuses by locally enforcing a high current, or using a laser etc. Alternatively, some sort of programmable/erasable memory (e.g. EPROM, or EEPROM etc. or, again, an array of fuses that can be selectively blown up) can be integrated on the die, and the status of the bits in this memory could drive the switches selecting the desired trim. In either case, once again, the idea is that the desired trim is not lost over time or if the device is turned off. More sophisticated trims include, for example, physically going over the device with a laser and selectively blowing up parts of it to alter its value.

The above discussion is not limited, as perhaps implied by the discussion so far, to physical components (such as resistors or capacitors, or even active devices such as a special BJT fabricated with multiple emitters etc.). It can be easily extended to a variety of other parameters such as the gain of a feedback amplifier (e.g. if that is set by the ratio of trimmed passive devices), the offset of a differential pair (e.g. by reducing the lithography-dependent contribution to the mismatch of the pair), or the temperature linearity of a PTAT (proportional to absolute temperature) current reference [93, 94, 95, 96] (e.g. the net temperature coefficient of a resistor could be reduced by building it as the series of two resistors with opposite-sign temperature coefficients;

however, the contribution of one resistor versus the other would need to be controlled), to improve the accuracy of data converters [52, 53, 97] and basically anything else that can be either directly or indirectly set as exemplified above.

The main, and significant, downside of trimming is its potentially high cost. The largest contributor to that is not necessarily the area on the die for the trimmable devices and the memory, but is often the time required to complete the trimming. Since that is performed during production testing on automatic test equipment (ATE), the time taken to reliably complete the trim can significantly add to the overall test time for the device, which can amount to a substantial percentage of the production costs of the individual device.

Trimming, together with the already-mentioned possible need to re-correct the device for changes that occur during operation, leads us to calibration. The general idea is, as stated above, essentially the same, namely the critical parameter is measured and then corrected. However, there are important differences. First of all, calibration, unlike trimming, can be performed multiple times after the device has been fabricated and shipped to the final user. Therefore it can be used to compensate for changes that have occurred over time and bring the device operation back within the desired range. Second, when we talk about calibration we often refer to *self*-calibration, which means that, unlike trimming, the measurement–correction process is entirely performed within the device itself and does not require any special external equipment along with the ATE. Third, since it is internal to the device, there is the potential to complete the calibration process in a shorter time than would be the case if this measurement–correction task were to be performed by trimming (e.g. by avoiding the insertion time for deployment of external equipment etc.). As stated, this is a *potential*, and that is a critical point, because complex calibration algorithms can require quite some time as well, and, since it is often required that the device be completely calibrated before ATE testing is completed, it can, again, lead to the same cost issue as has already been mentioned in the case of factory trimming.

A further distinction in the context of calibration refers to when that is performed. In particular, a data converter can be calibrated *while* it is performing the regular conversion operation, and the result of the conversion must not be affected by the calibration process itself. This is what is referred to as *background* calibration since the measurement–correction process is happening in the background and invisible to the outside world [98, 99].

Alternatively, the operation of the converter can be interrupted, whereupon the converter becomes unavailable to perform the normal conversion process, the calibration then takes place, and, once that has been completed, the converter again becomes available for its normal mode of operation. In the latter case we talk about *foreground* calibration [73, 100].

Expectably, there are advantages and disadvantages with each of the two approaches that need to be carefully assessed when devising self-calibration. Clearly, background calibration allows the converter to be used at all times and it is, in principle, completely transparent to the final user. This is crucial in applications for which the converter needs to operate on an uninterrupted time-series or when the critical data/sample arrives at a time instant that is not easily predictable and the conversion needs to take place immediately. Moreover, if the operation of the background calibration is faster than any correctable changes that may occur during the operation of the device then the calibrated converter is insensitive to the effect of these changes (it is said that the calibration "tracks" the variations). That is often true, for example, in the cases of temperature changes and average supply changes,[8] since generally these happen with relatively slow time constants compared with the calibration speed.

There are, however, important issues with background calibration that need to be accounted for. First, in order for parameter measurement to take place, in most cases, test stimuli (test signals, which can be either analog signals or digital data streams, depending on the specifics of the calibration taking place) need to be injected at critical nodes/sections, and, in fact, often these stimuli are superimposed on the converted (analog or digital) signal itself [99]. This signal superposition results in a larger signal range that the converter needs to accommodate. Second, some background algorithms require some units of the converter to be disconnected from the regular signal path to be calibrated, while an identical, pre-calibrated unit is connected to the signal path as a temporary replacement. Once the disconnected unit has been fully calibrated, that is re-connected to the signal path and a new unit is then taken off line for calibration in the same fashion. The process proceeds cyclically in the background on an ongoing basis. Although the unit calibration/rotation process should be transparent to the converted signal, it is not surprising that it can actually interfere with it, hence coupling undesired periodic disturbances to the converted signal. This coupling needs to be controlled or masked to be negligible.

On the other hand, during foreground calibration the converter is not processing any input signals. Therefore neither the signal range issue nor the coupling issue exists. However, the calibration is now unable to track directly environmental changes such as varying temperature or supply. To attempt to amend that, if the application allows it, the converter could be switched from regular conversion mode to calibration mode to be re-calibrated and to compensate for the changes that occurred while it was operating. In fact, similarly to what has previously been explained for background calibration, in some applications a number of foreground calibrated converters is employed in parallel, and, periodically, one will be taken off line for calibration and replaced by a pre-calibrated spare converter. In other cases, the application may allow time for taking the converter off line for calibration before it is used again (e.g. in certain communication applications data arrives in "packets" and there is no activity for some time between two subsequent packets). Finally, in other cases, the main reason for calibrating may be simply process variation. In this case foreground calibration may be performed once only, when the converter is powered up (and possibly after it has reached its thermal steady state), hence becoming available (and calibrated) for regular operation after that, similarly, in a way, to a factory trimming performed from scratch any time the device is turned on [73].

The last distinction we will consider in this section refers to *analog* and *digital* calibration. As the adjective suggests, in analog calibration the corrective action happens

---

[8] While some of the effects of slowly varying supply fluctuations might be reduced, supply *noise* cannot be corrected by calibration because it has a significant power content at very high frequency.

in the analog domain. So, for example, if, due to process-supply voltage–temperature (PVT) variation, a certain current drops in a section of an amplifier, hence leading to a reduction in bandwidth, then a compensation current may be injected to counter this variation and restore the desired bandwidth value. Adjusting the value of a passive component (similarly to the example of Fig. 2.11) by switching on/off additional components is another example of analog calibration.

When, however, the estimation of the critical parameters and the correction are applied in the digital domain (hence, after digitization in the case of an ADC, or before analog synthesis in the case of a DAC) then we talk about digital calibration [99, 101]. So, for example, if the ADC or the DAC suffered from a static nonlinearity, a (digital) model[9] for this nonlinear function $f(x)$ could be created and its parameters estimated. Then, an inverse function $f^{-1}(y)$ is digitally created and applied to the data stream at the output of the ADC or to that at the input of the DAC in order to cancel out the nonlinearity itself: $f^{-1}(f(x)) = x$. The purpose of the calibration, in this example, is to determine the proper $f^{-1}(y)$. Clearly, having a mix of analog and digital estimation and calibration can be optimal in many cases as well.

Before completing this section we will briefly examine two important considerations associated with calibration in general. First of all, the use of calibration can be exploited to design faster, and possibly smaller, circuits. For example, if calibration can take care of correcting for mismatch of devices or for parameter spread then many circuits can be designed/optimized for much higher speed, selectively and appropriately using smaller devices with associated smaller parasitics. Second, as we have seen in Section 1.2.1, as CMOS technology continues scaling down toward the nanometer range, analog designers face a number of important challenges in device matching, parameter spread, and supply voltage. Some of these challenges can effectively be tackled by means of suitable calibration techniques. Incidentally, if most of the calibration is performed in the digital domain, scaled CMOS processes are an ideal match to these techniques due to the reduced cost per digital function and the massive availability of transistors to be used for digital design.

2.6   Summary

This chapter was simply meant to give a refresher and, perhaps, to provide a slightly different view of many background concepts that will be needed in support of the material covered in the next chapters.

In Section 2.1 some of the performance metrics used in data conversion have been discussed. Although resolution and sampling rate are, without doubt, the two key parameters stated when quoting the specifications of a data converter, they are often insufficient to assess whether or not a converter will meet the needs of an application. Other metrics have been briefly and informally summarized.

---

[9] This could be a known closed-form model of the nonlinearity, a polynomial expansion, or a look-up table; there is a wealth of theory of "modeling and parameter estimation" behind this [102, 103, 104].

Following the engineering metrics of the previous section, various figures of merit or factors of merit (FOMs) have been discussed in Section 2.3. FOMs too, partly due to their extreme simplicity, don't often do proper justice to the performance of converters. Nevertheless, and in spite of the endless and animated debates they generate among practitioners, FOMs can be used as a tool to spot trends or to make some quick comparisons.

Empirical charts sketching where classic ADC and DAC architectures fit in the BW–SNDR and BW–SFDR spaces have been briefly discussed in Section 2.4. These charts open up many important questions. For example, how can we push the speed–resolution boundary further? Would that be possible by designing properly devised hybrid architectures (e.g. a pipeline-$\Delta\Sigma$ ADC or a current-steering-$\Delta\Sigma$ DAC) that take advantage of each architecture's strengths and compensate for the respective limitations? Is process technology driving the adoption and advancement of some architectures as opposed to others or, even, leading to the introduction of new architectures?

Finally, Section 2.5 briefly summarized the key concepts behind trimming and calibration. This is a huge topic that would require more than one entire book to cover properly; moreover, due to both the demand for higher-accuracy converters and the increasing challenges offered by nanometer-scale process technologies, the recourse to calibration for data converters is becoming a recurring necessity and it is thus driving a rapid expansion of this topic. Many of the advanced ADCs and DACs discussed in the following chapters will provide effective examples of smart application of new calibration techniques.

3 Advanced analog-to-digital converters

The first two chapters of this book set the stage for the following ones. This chapter covers some of the most advanced and recent architectures and circuit techniques for analog-to-digital conversion.

Certain classic ADC architectures have recently seen renewed interest and, in fact, have gained something like a "new life" when researchers and designers rediscovered them as very valuable options to push the converters' performance, especially as deep CMOS scaling begins to offer serious challenges to switch capacitor ADCs using precision amplifiers. The renaissance of these "forgotten" ADCs is the topic of Section 3.1.

Progress in analog-to-digital conversion, however, has certainly not relied only on rediscovering and modernizing old tools and techniques. A few emerging architectures and techniques that certainly break away from tradition to a larger degree are discussed in Section 3.2. Most of these, however, are still somewhat in their "incubation" stage since to date there aren't many commercial stand-alone ADCs based on them. This section offers a look at what the imminent future of ADCs could be.

Another recent trend in this field has been associated with the proliferation of hybrid architectures, namely ADCs whose architecture is not a single classical one (e.g. a pipelined ADC, a $\Delta\Sigma$ ADC, a flash ADC etc.) but rather a combination of two or more into a single converter. The idea is that, on carefully combining two or more architectures, the resulting one would have the advantages of both and then achieve better overall performance than if either one or the other had been adopted. This is the topic of Section 3.3.

Time-interleaved ADCs have been known for decades but it is only recently that they have gained something of a "mainstream" status among medium-resolution integrated ADCs thanks to significant advances in calibration. Section 3.4 will review some of the relevant concepts and highlight some examples of recent interleaved ADCs.

Examples of pipelined ADCs are described throughout this chapter in order to illustrate some new techniques. Moreover, the presence of this architecture is clearly felt in various sections through comparisons with other types of ADCs. A chapter on advanced ADCs could not be completed without devoting some words to one of the most pervasive architectures of our times. So, what about pipelined ADCs? Well, Section 3.5 concludes the chapter with a very brief digression on some recent results on the "good ol' pipe."

To clarify the notation in the following sections, n represents the resolution or "number of bits," f_s is the sampling rate, V_{in} is the continuous-value/continuous-time input signal,

BW is the input signal bandwidth, $\hat{V}_{in}$ is the continuous-value/discrete-time *sampled* input signal, and $\tilde{V}_{in}$ is the latter signal after quantization. Finally OSR $= f_s/(2 \cdot \text{BW})$ is the oversampling ratio and $D_{out} = (d_{n-1}, \ldots, d_1, d_0)'$ is the digital output in binary (vector) representation.

3.1   The renaissance of some classic ADCs

Most of the nineties and the early part of the first decade of the new century saw the "golden age" of the switch capacitor technique. This extremely powerful analog signal processing technique was employed to design virtually all types of analog circuits, including filters, multipliers, and, of course, ADCs and DACs [5, 33, 105].

That was enabled by the availability of CMOS processes with supply voltages large enough to offer the necessary headroom for the design of high-gain/high-bandwidth amplifiers and high-quality switches (high off-impedance, low on-resistance, large signal compliance, and reasonably small stray capacitances). Furthermore, the emergence of independent foundries such as TSMC, UMC, and CSM, making such process technologies easily accessible and relatively affordable, enabled the fab-less IC design business model and hence the birth of countless IC design start-up companies. Some of these evolved into larger companies and now stand side by side with the "historical" semiconductor corporations that always relied on internal foundries as their foundation.

The above dynamics fueled a very broad and diverse range of experimentation and innovation. The simultaneous and independent emergence of the pipelined ADC architecture, a perfect match to the switch capacitor technique, turned out to be an "attractor" of many of these innovation efforts. This architecture was faster than the successive approximation (SAR) ADC (which, at that time, was mostly used in a variety of general-purpose and industrial automation applications) and of higher resolution and lower power than the flash ADC (which was then popular in disk drive applications and optical applications). It also happened to be a very good match in bandwidth and dynamic range for a large variety of applications, in particular for the then rapidly growing and profitable wireless infrastructure market.

This explosive convergence of process and circuit technologies, together with new business models and a flourishing world economy, literally created a condition in the data conversion field whereby pipelined ADCs unequivocally became synonymous with the present and future of analog-to-digital conversion, relegating $\Delta\Sigma$ ADCs to precision and audio niches and, perhaps, pointing all other architectures (including SAR ADCs) toward a gradual, but seemingly inevitable, retirement.

However, this dynamic began to change. As discussed in Section 1.2, the relentless digitally driven march of Moore's law toward more and more scaled processes and the associated lowering of supply voltages, together with the rapidly increasing process costs, the dot-com boom, and mutated economic conditions, reversed all the conditions which had led to that, perhaps somewhat irrational, euphoria and to the unquestioned predominance of pipelines.

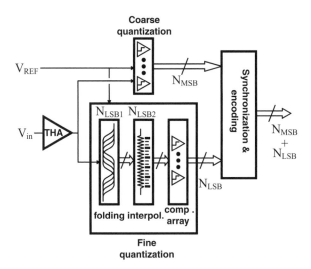

Figure 3.1 Block diagram of a traditional folding and interpolating ADC. Two key blocks can be recognized: a coarse quantizer (a flash ADC) that identifies the fold within which V_{in} lies and a fine quantizer (a combination of a folding circuit, interpolation, and quantization) that quantizes V_{in} within one of the input range folds. Coarse and fine quantization are then combined by a synchronization and encoding block to represent the final output. The numbers of bits resolved (not necessarily the number of lines among the blocks) are progressively indicated as N_{MSB} (for the coarse quantization) and $N_{LSB} = N_{LSB1} + N_{LSB2}$ (fine quantization: N_{LSB1} from folding and an additional N_{LSB2} increasing the resolution by interpolating more zero crossings). The coarse and fine quantizers need to be finely aligned in voltage and, possibly, in time.

ADC designers had to find other options and, needless to say, the first and most natural recourse was to take another look at the old alternatives and to wonder whether, with a proper update, these might be, once again, valuable tools with which to perform the desired data conversion. It is because of this that in recent years we have witnessed something of a "renaissance" of classic ADC architectures, including SAR ADCs, folding and interpolation ADCs, and continuous-time $\Delta\Sigma$ ADCs. Some people are now even wondering whether the recurrence of the very same architectures as had originally been conceived decades ago isn't perhaps an indication that these are, in fact, fundamental in nature [106].

3.1.1   Folding and interpolation ADCs

The folding and interpolation ADC (FI ADC) is one of the "classic" ADC architectures which have been around for a long time [107], and it has been described in textbooks such as [34, 52, 53].

Briefly recapitulating the key concepts, such an architecture is composed of a *fine* and a *coarse* quantizer as sketched in the high-level block diagram of Fig. 3.1. The full analog input range $[V_{hi}, V_{lo}]$ that can be digitized is partitioned into intervals of equal width, Δ: $[V_{hi}, V_{hi} - \Delta], [V_{hi} - \Delta, V_{hi} - 2 \cdot \Delta], \ldots, [V_{lo} + \Delta, V_{lo}]$.

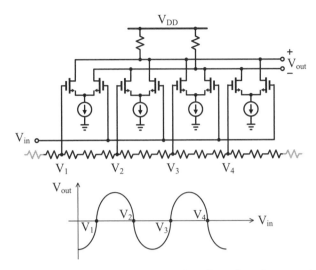

Figure 3.2 The concept of folding using four differential pairs and a reference resistive ladder.

The coarse quantizer determines within which of these sub-intervals the input signal V_{in} resides.

The fine quantizer senses the whole input range $[V_{hi}, V_{lo}]$ but it "folds" it into the previously mentioned intervals of width Δ. Namely, it determines (quantizes) where V_{in} is located within one of the Δ intervals but it cannot determine whether that is, for example, the interval $[V_{hi}, V_{hi} - \Delta]$ or the interval $[V_{lo} + \Delta, V_{lo}]$ or any one of the others.

It is the combination of the information between the coarse and the fine quantizer that provides the final digitized output. A good analogy to that is the case of a traditional clock, where the time reading is obtained by combining the information from the hours arm (the coarse quantization) and the minutes arm (the fine quantization) [108]; the coarse quantizer alone tells us only which one-hour interval the present time is within, while the fine quantizer cannot distinguish between one-hour partitions but, within the present hour, determines what minute it is.

While the coarse quantizer is usually a relatively low-bit-count flash ADC, the fine quantizer makes use of three analog techniques to finely discriminate the position of V_{in} within the Δ intervals.

- *Folding* [107], see Fig. 3.2. The input range of a differential pair, pre-amplifying the difference between the input signal and a reference level, ahead of feeding it to a comparator, is limited; so multiple input ranges, corresponding to multiple such pairs, are merged together and provided to a single comparator for zero-crossing detection. These ranges are then indistinguishable by the comparator, hence being "folded" into one another. By doing this, it is possible to reduce the number of comparators required.
- *Interpolation* [107], see Fig. 3.3. The outputs of multiple differential pairs, each one amplifying the difference between the input signal and an (explicit) reference level, can also be used to determine the proximity of the input to additional (interpolated)

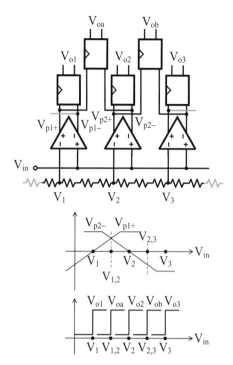

Figure 3.3 The interpolation concept: a virtual zero crossing is detected/interpolated in between two actual reference points by suitably feeding the inputs of an extra latch with the outputs of the pre-amplifiers sensing V_{in} and the reference voltages. Interpolation can also be implemented by inserting a resistive (or capacitive) divider between adjacent pre-amp outputs (similarly to averaging networks) and then tapping off the intermediate levels from that to be detected. Adapted from B. Razavi, *Principles of Data Conversion System Design*, IEEE Press, 1995.

reference levels. By doing this, it is possible to reduce the number of pre-amplifiers for a given set of reference points, or, equivalently, it is possible to increase the number of reference points (resolution) for a given set of pre-amps.

- *Averaging* [109, 110], see Fig. 3.4. The linearity of the conversion can be improved by averaging the effects of the mismatches in the analog circuitry involved by introducing suitable mutual coupling between adjacent elements.

It should be noticed, in passing, that all of the above three techniques, although critical for an FI ADC's implementation, aren't completely exclusive to FI ADCs, and have been used in multiple other architectures. In fact, while folding and interpolation were developed for FI ADCs, averaging was originally introduced for a flash ADC [109].

The FI ADC was another architecture that during the late nineties and early 2000s, particularly in CMOS processes, seemed at risk of being "smashed" between the super-fast flash ADC and the increasingly popular pipelined ADC. Flash ADCs with giga-sample per second (GSPS) sampling rates held firm with resolution up to 6–8 bits [111, 112, 113]. Pipelined ADCs were rapidly gaining ground with increasingly higher sample rate (hundreds of mega-samples per second, MSPS, for single ADCs [114] but

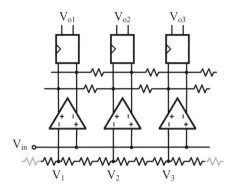

Figure 3.4 The averaging concept: mutual coupling between adjacent pre-amps introduces spatial averaging/filtering of the offsets, hence improving the DNL [110].

a few hundred MSPS to GSPS for the emerging time-interleaved ones [115, 116, 117]) and resolution as low as 8–10 bits (though the ENOB performance of such GSPS ADCs was more in the $\sim$6 b range).

At that time the difficult challenges to the FI ADCs' performance were primarily the following.

- Linearity (and hence also resolution) was primarily limited by device matching, specifically, the ability to control the position of the zero-crossing points in the internal FI stages of the fine quantizer as a result of mismatches. As briefly recalled in Section 1.2.1 and Eqs. (1.1) and (1.2) (and more extensively in Section 4.2.2 later in this book), higher intrinsic MOS device matching can be obtained at the expense of higher overdrive voltage $V_{GT} = V_{GS} - V_T$ (leading to voltage headroom limitations) and higher device area $W \cdot L$ (leading overall to larger parasitics as well).
- In addition to speed, the higher transconductance g_m of BJTs, compared with MOS transistors, also contributed to better control of the zero-crossing points [118] and, in turn, to some extent suggested that bipolar/BiCMOS technologies would be a better process fit than standard CMOS.
- Larger and parasitic-heavy circuits inevitably become plagued by slow time constants and can be quite power hungry, making them inefficient and unattractive.

Note also that the inability to properly control the position of the zero crossings as a result of mismatch, process, temperature, and supply variations (PVT variations), together with the practical available active range of the differential pairs in the folders (only one needs to be in the linear range for a given input level, while the others must be fully saturated, either to the positive or the negative full scale), limits the number of foldings that can be reliably implemented in a single stage.

As a result of all of the above, although FI ADCs were still more attractive than flash ADCs with respect to power efficiency and input impedance, and still considerably faster than many (non-time-interleaved) pipelined ADCs, these gaps were narrowing, while it appeared to be unpractical to use FI ADCs for more than 6 b and applications faster than a few hundred MSPS.

Some of the important breakthrough works that challenged all that and brought FI ADCs renewed attention were the 2004 papers by Taft *et al.* [73] and Geelen and Paulus [119], respectively describing a low-voltage 8 b 1.6 GSPS FI ADC and an 8 b 600 MSPS FI ADC, both in 0.18 μm CMOS processes.[1]

As described in far greater detail in [73] and [120], the performance of the former ADC was achieved by successfully applying a number of techniques, some of which will be mentioned in the following.

First of all, the issue of displaced zero-crossing points in the folding stages, resulting from offsets in their MOS differential pairs, was primarily addressed by means of foreground calibration. By doing this it was possible to design such circuits for the highest possible speed without worrying about matching.

Also, while in a traditional FI ADC with n bits (2^n zero crossings) there is a trade-off among the number of foldings F_F (also known as the *folding factor*), the number of interpolations F_I (also known as the *interpolation factor*), and the number of repetitions of the primary folding function N_P [118],

$$n = \log_2(N_P \cdot F_F \cdot F_I) \tag{3.1}$$

this limiting element was circumvented and, consequently, the complexity of the FI stage itself was reduced, by pipelining subsequent FI stages [108, 121, 122].

Furthermore, the whole ADC was actually a time-interleaved (TI) ADC composed of two TI FI pipelined sub-ADCs, each one working at half the sample rate (800 MSPS). Both ADCs were independently foreground calibrated so that also their gain and offset mismatches were low enough to make the corresponding interleaving spurious spectral content negligible.

The two TI ADCs each had an independent linearized open-loop track-and-hold amplifier/stage (THA). The sampling switch was linearized by a constant-V_{GS} bootstrapping while the static linearization of the THA stage was also included in the calibration of the full ADC [73, 120]. Timing skew mismatch between the two THAs was sufficiently addressed by gating the sampling instant from a full-rate master clock as well as by careful circuit and layout design.

As mentioned before, the renewed interest in FI ADCs resulted in further advances. For instance, besides what has just been described, an additional benefit of introducing calibration (removing, for example, the large device-area constraints introduced by matching requirements) is to give the designer the opportunity for possible power reduction (see also Section 2.5 and [29], where an explicit tie between matching and power consumption is discussed). So, alongside that and in contrast with the foreground calibration approach of [73], Nakajima *et al.* [123, 124] proposed a background calibration algorithm that also calibrates the comparators for a somewhat similar 6 b 2.7 GSPS FI ADC.

---

[1]   Although the converter performance is what really matters, it is interesting to note that performing a search on the IEEE/IET archive IEEEXplore will return only around 10 publications on this topic between 1999 and 2004, while the tally jumps to more than 35 between 2005 and 2010.

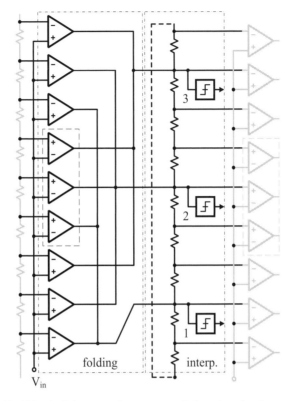

Figure 3.5 The unified FI principle: a set of comparators is introduced at the output of the folding stages to signal which fold is active, hence merging the coarse quantization with the fine one [129]. © 2009 IEEE.

Instead, a digital calibration for an 8-bit FI ADC incorporating redundancy and reassignment was described in [125].

Furthermore, power consumption continued to be a challenge. Some power-saving techniques were proposed for folding and interpolating blocks in [126].

However, again, two of the main barriers to further increasing both speed and resolution with FI ADCs remained the active device transconductance g_m and the ability to match, average, or calibrate the circuit elements determining the zero-crossing points. So, while in [127] the higher resolution of 10 b (at 1 GSPS) was also achieved in an FI ADC by taking advantage of bipolars, the FI ADC in [128] achieved 10 b resolution at a sample rate of 200 MSPS, while staying with a 0.13 μm CMOS, by leveraging an open-loop autozero technique to cancel out pre-amplifier offsets and allowing the pre-amplifiers to provide sufficient gain to overcome offsets from the following stages.

In 2009 another significant advance was published by Taft *et al.* [129]. A 10 b 1 GSPS FI ADC on a 0.18 μm CMOS process using a new "unified folding" architecture merging the coarse and fine quantizer and, in addition, considerably pushing both resolution and sample rate for a CMOS converter was presented.

A conceptual diagram illustrating a key element of this unified FI architecture is depicted in Fig. 3.5. The ADC is composed of a cascade of FI stages (one of which is

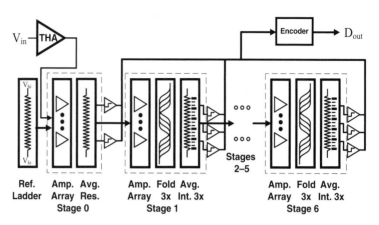

Figure 3.6 Unified FI ADC block diagram [129]. © 2009 IEEE.

depicted in Fig. 3.5). Each stage contains classic FI circuitry partitioning and folding the input range into sub-ranges and interpolating zero crossings. However, comparators are added to the stage in order to detect within which fold the input of the stage lies. This can also be interpreted as resolving some of the higher-order bits and then passing the analog output of the FI stage to the next one for a finer quantization.

That resembles the principle of operation of a pipelined ADC, with the immediate difference, however, that in the unified FI ADC the conversion simply ripples through the open-loop stages and does not require DACs and clocked residue generation (the latter being gated by a quantization/decision of the stage input) [129].

The merging of the coarse and fine quantization allowed a very large folding order (overall $3^6 = 729$), which in turn was obtained by cascading the FI stages. In fact, the particular choice of the folding factor ($F_F = 3$), interpolation factor ($F_I = 3$), and number of comparators per stage (3) led to a modular and recursive structure of 7 stages that used only 20 comparators (in fact 2 for the first stage and 3 for each of the subsequent 6 stages) to detect 2187 levels, namely sufficient for 11 bits. The encoder output, however, was truncated to 1025 levels, namely 10 b plus over-range [129]. This 10 b ADC, like the prior 8 b ADC, is also foreground calibrated, and it is also composed of two TI ADCs, each one working at half the sample rate.

A block diagram of the ADC is shown in Fig. 3.6.

3.1.2   Dynamic ADCs

Many traditional ADCs necessitate class A or AB circuits as building blocks. As a result, during normal operation, they drain substantial DC supply current irrespective of the input signal V_{in}, and, in fact, the overall power consumption can be dominated by static power consumption. Examples of that are many traditional pipelined ADCs, folding and interpolating ADCs, $\Delta\Sigma$ ADCs etc. In some of these examples, the need for large static current in some of the circuit blocks results from the need to meet distortion and noise specifications.

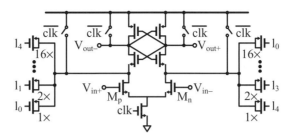

Figure 3.7 A dynamic comparator with built-in reference and trimmable offset [130]. © 2006 IEEE.

On the other hand, for example, in successive-approximation converters (SAR ADCs) the input signal determines the actual step-by-step operation of the converter, and therefore it can be expected that the dynamic power dissipation is a non-negligible part of the overall power consumption; in fact, it generally increases with the frequency and amplitude of V_{in}.

The ever increasing need for power-efficient converters, particularly in such applications as mobile applications, medical implants, sensor applications etc., has provided significant motivation to develop converters with the lowest possible static power dissipation, namely ones for which the overall power dissipation is a strong function of the input signal activity.[2]

In order to achieve that, researchers have explored very different options.

In some cases the sampling process has been kept uniform in time. But the operation of the ADC is such that the amount of charge drawn from the supply at each clock cycle is strongly dependent on the internal conversion process and, hence, on the converted signal V_{in} [130].

Conversely, in other cases, the sampling and conversion of the input V_{in} happens only when the input experiences sufficient variation for it to be deemed worthy of being acquired [131, 132, 133].

In [130] a 4 b 1.25 GSPS flash-like ADC wherein the static power consumption is systematically minimized has been described. All blocks that traditionally lead to static current draw have been eliminated and their functions have been implemented in alternate ways. Specifically, these blocks are the THA, the pre-amplifiers, the reference ladder, and the bubble-error-correction circuitry.

Since the reference ladder is eliminated, the reference levels have been built into the comparator by intentionally making the devices of each comparator's input pair different in size. This dynamic comparator is shown in Fig. 3.7 and the widths of the input devices M_p and M_n are set as $W_p = W + i \cdot \Delta W/2$ and $W_n = W - i \cdot \Delta W/2$, respectively (with $i = -7, \ldots, 7$).

---

[2] This should not be confused with a different trend, that of adaptive ADCs, where the converter's mode of operation is changed (by enabling/disabling blocks or changing their quiescent point) as a result of the conversion needs (the nature of V_{in} or the corresponding conversion specifications). In the latter category the power consumption is still primarily dominated by static dissipation, with the difference that this is minimized by the mode of operation for the desired mode.

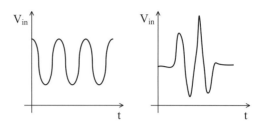

Figure 3.8 A periodic signal (left) versus an aperiodic signal (right).

To minimize power further, the size of the comparator's devices is made as small as possible. The resulting mismatches/offsets are foreground calibrated. The correction is accomplished by varying the comparator's output load by means of the small binary weighted PMOS capacitors depicted in Fig. 3.7. These PMOS capacitors effectively load the comparator's output when their gates are tied to ground ($l_n = 0$ V) and charge is accumulated in their channels, while their capacitive load to the output is minimal when their gates are brought to the upper supply ($l_n = 1.2$ V) and the channel is not formed. Once each comparator's trip point has been calibrated by properly setting the bits $l_0, \ldots, l_4$, the flash ADC is ready to be used.

The 15 comparators directly sample the input (hence eliminating the THA) and their thermometric output is captured and held by 15 set–re-set latches. The outputs of the latter are used to drive a ROM-based 4 b Gray encoder with intrinsic error-correcting properties.

So, again, static power has been minimized by systematic removal and replacement of the blocks that traditionally sink most of the DC supply current.

A very different approach, called "event-driven" data conversion [133, 134, 135], originating from the same goal of minimizing the energy utilized to digitize a continuous-time signal $V_{in}(t)$, is suitable for a wide class of essentially non-periodic signals. In many common applications such as audio, particularly speech processing, medical applications, particularly ultrasound diagnostic or biomedical monitoring, disk readers, seismic activity monitoring, radar applications, and several other cases, the continuous-time signals being processed are far from being periodic. They commonly feature time intervals during which there is little or no variation ("silence") followed by limited time intervals during which the signal experiences large activity (possibly with varying frequency content) before returning to a "quieter" state [136] as exemplified in Fig. 3.8.

A traditional periodic (synchronous) Nyquist sampling of signals of this nature can be quite wasteful in terms of energy. That is because f_s needs to be at least twice the bandwidth of $V_{in}(t)$, and hence the ADC samples the periods of "silence" or low-frequency content at the same rate (hence, possibly, with the same consumption) as it does the rare periods of fast variation ("bursts"). It is, instead, considerably more efficient to adapt the sampling activity to the rate of variation of $V_{in}(t)$: slower sampling rate when $V_{in}(t)$ is quiet and faster when it is more active.[3]

---

[3] That is essentially the same as what is done in adaptive step-size integration for analog transistor-level simulators or, hence the denomination, in "event-driven" digital simulators [137, 138].

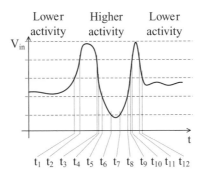

Figure 3.9 Level-crossing sampling. The quantization levels are marked by horizontal dashed lines. When V_{in} crosses one of these levels, a sample $(t_k, V_{in}(t_k))$ is captured. It is possible to distinguish periods of time when the signal has lower activity and fewer samples are taken, while a higher sample rate arises as a result of higher activity [142]. © 2011 IEEE.

With this type of asynchronous sampling, since the sampling instants t_k can no longer easily be identified as $t_k = kT_s$ ($k \in \mathbb{N}$), samples are pairs of sampling time and input level $(t_k, V_{in}(t_k))$ ($t_k \in \mathbb{R}$). The question now becomes when does the signal $V_{in}(t)$ need to be sampled? Different approaches have been proposed. The oldest, simplest, and most common is the "level-crossing" sampling whereby $V_{in}(t)$ is captured when its value is equal to specific reference levels [131, 132] as depicted in Fig. 3.9. Alternate approaches monitor the rate of change of the signal. For example, samples are captured when the absolute difference between the present input value and the prior one is larger than a prescribed amount [139, 140]. Alternatively, the slope of the input signal is estimated and a sample is taken when that crosses a defined threshold. It is also possible, where the local sampling rate is high, to eliminate intermediate samples and approximate them by means of piecewise amplitude reconstruction or other "signal-compression" techniques [133, 141].

In addition to the potential for large power saving, this type of asynchronous operation has other important theoretical advantages over traditional synchronous sampling ADCs. For example, in traditional synchronous sampling the input signal harmonics and the quantization noise outside the Nyquist band are aliased back to the Nyquist band, hence limiting the SNDR to its theoretical maximum stated by the relation 6.02ENOB + 1.76 dB (see Eq. (2.1)). However, since with asynchronous sampling the amplitude samples are uniquely identified together with their actual sampling time, aliasing does not occur [133] and the previous limitation on quantization is eliminated. For example, more than 8 effective bits have been obtained using 3–6-bit quantizers [133, 142].

The most straightforward implementation of a level-crossing quantizer probably consists of a flash ADC without a sampling clock [143]. However, a more hardware-efficient implementation is the continuous-time ADC depicted in Fig. 3.10 [144, 145]. In this scheme, the input $V_{in}(t)$ is duplicated and biased up and down at INH and INL, respectively, to be sensed by the two comparators. The second input of each comparator

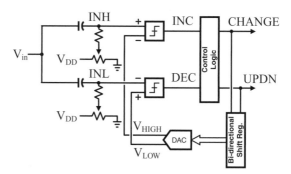

Figure 3.10 A continuous-time Δ ADC [144]. © 2008 IEEE.

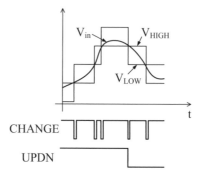

Figure 3.11 As the input signal $V_{in}(t)$ varies the reference levels V_{HIGH} and V_{LOW} are updated to track it. Accordingly the digital outputs of the Δ loop CHANGE and UPDN signal that the input has crossed into a new quantum and specify its direction of change.

is a voltage, $V_{HIGH}(t)$ and $V_{LOW}(t)$, respectively, generated by the feedback DAC. The operation of this architecture can be explained by looking at the waveforms depicted in Fig. 3.11.

Let's assume that the input $V_{in}(t)$ lies between the present values of $V_{HIGH}(t)$ and $V_{LOW}(t)$. In this case both comparators have zero output and the digital outputs CHANGE and UPDN don't change state. As $V_{in}(t)$ varies, it does at some point cross one of the two DACs' output voltages. So, for example, if it increases then it will cross $V_{HIGH}(t)$. Then INC becomes 1 (while DEC stays at 0), and the control logic delivers a short pulse at CHANGE, signaling that a threshold has been crossed, while UPDN is set to 1 to indicate that the direction of change is upward. In other words, the digital output is communicating the crossing of a quantization level and the direction of change of the input. This information, together with knowledge of the prior state of $V_{in}(t)$, allows one to determine the digitized input level, which is used to update the DAC output in order to track $V_{in}(t)$. The use of a bi-directional shift register to control the feedback DAC simplifies its design and avoids digital glitches since the DAC's input always changes by one step at a time. This is an asynchronous Δ ADC and, as explained, its power consumption is a strong function of the input signal activity.

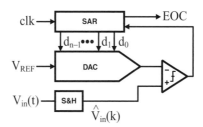

Figure 3.12 The block diagram for a traditional successive-approximation ADC.

Some observations are in order. If $V_{in}(t)$ changes rapidly and has a wide swing, then it will rapidly cross several quantization levels. Since, regardless of the implementation, it takes a finite amount of time to capture a sample when a level is crossed, a signal of this type can be subject to distortion since the ADC might not be sufficiently fast to properly acquire all the samples. This is akin to "slew-rate-limiting" behavior when the sample rate attempts to go beyond a maximum sample frequency determined by the time required to process a single sample [142, 146]. It is intuitive to understand that this class of asynchronous ADCs suffers from an accuracy-versus-speed trade-off: smaller and slower signals can be digitized with higher accuracy while large and rapidly varying signals lose accuracy.

Similarly to what is done in signal processing with adaptive differential PCM and adaptive delta modulation, it is then possible to consider making the quantization granularity an adaptive function of the input signal dynamics, hence increasing the conversion efficiency further [146]. In order to accomplish that, the slope of the input signal $V_{in}(t)$ is monitored and the resolution of the quantizer is varied accordingly. In [146] it has been shown that a simple way to dynamically reduce resolution and gain speed is to skip an even number of samples before acquiring the next input. This simple approach requires minimal overhead and therefore allows the intended power saving without too much additional hardware (which would have required more power, hence defying the main goal).

A different approach is used in [142], using an ADC architecture similar to that of Fig. 3.10. In the latter, the LSB's size is dynamically changed using the DAC as the slope of $V_{in}(t)$ varies within predefined ranges. In this way, if $V_{in}(t)$'s slope increases considerably, the quantization steps widen (reducing resolution), leading to a reduction in level-crossing events and hence allowing more time for the sampling process.

3.1.3   Successive-approximation ADCs

One of the oldest and best-known ADC architectures is certainly the successive-approximation ADC (often referred to as SAR ADC, where SAR actually stands for successive-approximation register) [5, 147, 148, 149]. A generic block diagram of a traditional n-bit SAR ADC is depicted in Fig. 3.12.

The input $V_{in}(t)$ is first sampled by the sample and hold (S&H). This can typically take two or three clock cycles to allow accurate settling [149]. After that, the sampled input

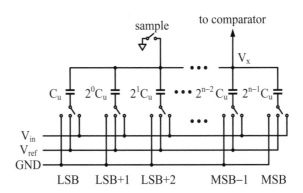

Figure 3.13 A traditional binary weighted charge redistribution network used to implement the S&H as well as the DAC function in successive-approximation ADCs.

is compared with the output of the DAC (which, in most modern implementations, is a capacitive charge redistribution DAC; in fact, the S&H function is often combined into the DAC array, as in the network depicted in Fig. 3.13, and it can easily be implemented in a differential fashion [150]) and a synchronous binary search determining the state of the n bits, beginning from the MSB and down to the LSB, changes the output of the DAC until it reaches $\hat{V}_{in}(k)$ to within less than an LSB, in n clock cycles [5, 149].

Once the input has been sampled, no additional processing occurs on it. No linear gain blocks are strictly needed (including in the S&H) for the SAR ADC operation. That is an extremely important and positive aspect considering the challenges in designing linear amplifiers in deep nanometer CMOS processes (see Section 1.2.1). On the other hand, the absence of gain blocks also implies that each of the n decision steps requires full accuracy [151].

The accuracy limitations of the SAR ADC are those of its key blocks.

- S&H accuracy is limited by the switches' noise, charge injection, clock feedthrough, aperture jitter, and settling time; $\hat{V}_{in}(k)$ is the representation of $V_{in}(t)$ so it must be at least as accurate as its desired digital conversion.
- The switch capacitor DAC output is compared with $\hat{V}_{in}(k)$, so its settled output must be as accurate as $\hat{V}_{in}(k)$; mismatches between the capacitors result in static linearity errors; at a high level, matching and segmentation considerations are analogous to those for current arrays in current-steering DACs discussed in Chapter 4 and can be addressed through layout techniques [47, 64], trimming, and/or calibration [100]; moreover, the DAC's noise contributes to the ADC's noise.
- The comparator: unless scaling is performed during the conversion process, the comparator's offset results in an ADC offset and does not impact linearity; its noise contributes to the ADC noise.

Similarly, the speed limitations come from

- the DAC settling, which is generally dominated by the settling of the reference charging the DAC's capacitors; roughly, in order to settle to within 1/2 LSB, the time constant

τ_{eq} associated with the charging process needs to be

$$\tau_{eq} < \frac{T}{(n+1) \cdot \ln 2} \qquad (3.2)$$

where T is the time allowed for the settling;
- the comparator decision time;
- the logic decision delay of the SAR controller driving the DAC.

The first two items generally dominate.

Insofar as the power consumption is concerned, a small percentage of it (about 10%–20%) is associated with driving the switches, the logic circuitry, and the comparator, while the largest percentage comes from the reference charging the capacitors in the DAC. That is particularly true for high resolution since the total capacitance of the array grows very rapidly with the required accuracy.[4] Another non-negligible source of power consumption which is often overlooked relates to the power supplied by the signal source ($V_{in}(t)$) to charge the sampling capacitor of the S&H.

Because of the above,[5] the power consumption of SAR ADCs tends to increase linearly with the sample rate f_s and, instantaneously, is a strong function of the converted input ($V_{in}(t)$) analogously to some of the considerations previously presented for some of the dynamic ADCs in Section 3.1.2.

Its natural amenability to CMOS process scaling, together with its conversion power efficiency, has led, during the last 5–10 years, to a phenomenal comeback in popularity of the SAR architecture. In particular, the latest SAR ADCs are finding a "sweet spot" in the space around 8–12 b resolution and 1–100 MSPS sample rate which used to be a pipelined-ADC-dominated space until only 5–10 years ago [153].

A passive-sampling and charge-bi-section 9 b/50 MSPS SAR ADC with the goal of maximum power efficiency has been introduced in [153]. The architecture is shown in Fig. 3.14.

The passive track and hold (T&H) is shown on the left, while the DAC shown on its right uses a binary weighted capacitor array in which all capacitors are precharged to the supply voltage (used as reference voltage, hence doing away with active reference buffers) before the conversion process begins.

An offset-calibrated dynamic comparator similar to the one shown in Fig. 3.7 is used and depicted on the right of the DAC array.

The whole conversion process occurs in the charge domain instead of the voltage domain, eliminating the need for active linear circuits and avoiding voltage-mode distortion due to nonlinear capacitive strays [154].

The input voltage $V_{in}(t)$ is applied to C_T through S_T. Then it is sampled into C_S (charge sharing between C_S and C_T occurs) and held there for the conversion, freeing up C_T to acquire the next input while the conversion process happens. The following

---

[4] Unless "double-array" structures are used, the total capacitance of the DAC array is $C_{tot} = 2^n C_u$, where C_u is the unit capacitance chosen on the basis of matching and/or $k_B T/C$ noise considerations commensurate with the desired accuracy. Use of a double-array structure relaxes the total capacitance by a factor of 2 [152].

[5] Again, assuming that no gain blocks are used in the S&H or for reference buffers.

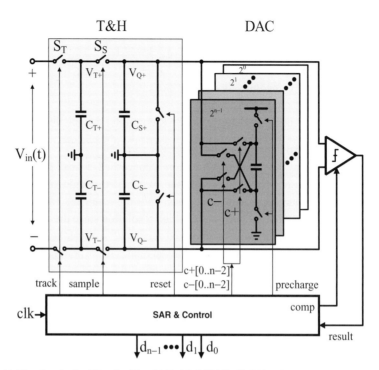

Figure 3.14 The Craninckx–Van der Plas SAR ADC [153]. © 2007 IEEE.

conversion process is essentially the traditional SAR binary search, whereby the binary weighted capacitors are sequentially shunted with C_S either with the positive polarity or inverted, depending on the sign of the comparison. An example of the waveforms involved is depicted in Fig. 3.15.

As the conversion progresses, determining the bits from the MSB toward the LSB, the subsequent charge-sharing steps drive the net charge left on C_S toward zero.

The binary weighted array DAC, in reality, requires one more technique for its practical implementation. Owing to the large size spread in capacitance between the DAC's MSB capacitor ($2^{n-1}C_u$) and the LSB capacitor (C_u), impractically small capacitances are required for the latter few bits. Therefore, an additional scaling is introduced by means of charge bi-section, for the lowest three LSBs as depicted (in single-ended fashion to reduce diagram clutter) in Fig. 3.16.

In other words, for these LSBs, lower charge is obtained on scaling down the precharge voltage as opposed to scaling down the storage capacitances. For this reason, the offset of the comparator would have introduced a conversion-dependent contribution to the process (resulting in distortion, instead of a mere conversion offset); that is the reason why an offset-zeroing scheme needs to be used.

This is an example of how offset and noise at the input of the comparator can introduce an error in the decision of the comparator. That is particularly true when the voltage difference between the two inputs of the comparator is very small. Such a difference, in

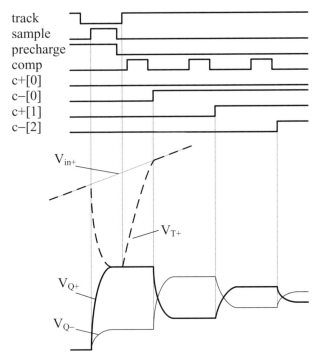

track
sample
precharge
comp
c+[0]
c−[0]
c+[1]
c−[2]

V_{in+}

V_{T+}

V_{Q+}

V_{Q-}

Figure 3.15 Waveforms for an example conversion [153]. © 2007 IEEE.

the case of an SAR ADC, would be small when the DAC output at the present step is very similar to the converted input $\hat{V}_{in}(k)$. Then random noise or offset superimposed on this small signal difference can be instantaneously large enough to change the sign of the net input voltage of the comparator, hence causing an erroneous bit decision.

This is actually a general issue for all ADCs, since every architecture uses comparators somewhere in the system, although its impact varies depending on the specifics of the architecture. Furthermore, when the input of the comparator is very small, another problem occurs: the comparator's reaction time can be very slow and its ability to make a decision (right or wrong) on time degrades considerably. This issue is known as *metastability* [34, 52], and it is particularly troublesome in very-high-speed and low-resolution converters, such as flash ADCs and folding and interpolation ADCs, where the time available for the decision is particularly short and the device noise and quantization noise are comparable in power.

Without digressing further on metastability and going back instead to the decision errors due to offset and noise, one way to mitigate the issue is through the introduction of *redundancy*. Redundancy is well known by pipelined ADC designers as a means by which to recover from offset errors in the flash sub-ADCs of the pipelined stages, by adding extra signal range, extra comparators, and so-called "digital error correction" to each pipelined stage. By adding all this additional hardware (redundancy, indeed) it is possible, as long as the comparator offsets are bounded to the designed redundancy,

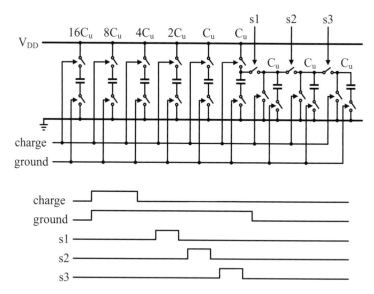

Figure 3.16 A capacitive DAC (single-ended only for convenience) with voltage bi-section re-scaled LSB segment and waveforms for the precharge sequence [153]. © 2007 IEEE.

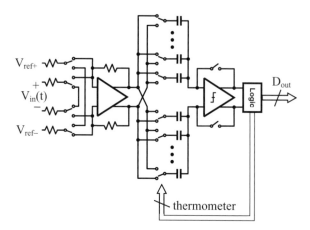

Figure 3.17 Kuttner's SAR with a thermometric capacitive array and 1.85 radix [151]. © 2002 IEEE.

to detect and recover from intermediate decision errors that could have led to severe linearity issues [155].

Redundancy in an SAR ADC has been described in [151] for a 10 b/20 MSPS ADC. The ADC is depicted in Fig. 3.17.

The capacitor array is segmented into a thermometric upper/MSB array made of 16×16 unit capacitors C_u and a binary weighted lower/LSB array for the remaining two bits. In addition to performing the DAC function during the conversion process, this array is also used to sample the input.

As expected, the input $V_{in}(t)$ is sampled first, while the comparator is closed into unity feedback so that its offset is sampled on the capacitors' plates tied to its inputs. During the following cycles, the reference is sampled on the capacitors according to the DAC code selected by the SAR control logic. Having previously sampled the input V_{in} minus the comparator offset, the comparator inputs are presented with the voltage difference between the latter and the DAC output. The SAR algorithmic conversion proceeds as usual; however, a key difference, redundancy, is introduced.

The redundancy is actually not explicit in the analog domain, but is instead built into the digital operation of the converter, as we are about to describe.

In a regular SAR ADC, during the first conversion cycle, the input V_{in} would be compared with $1/2$ of the scale V_{ref} of the DAC (in this 10 b case, that would correspond to code 512) to determine whether V_{in} is higher or lower than that. Then, depending on the comparator decision, the next cycle would change the DAC output to either $1/4$ of the scale or $3/4$ of the scale (i.e. respectively to code 256 or code 768) and so on.

However, a radix lower than 2 is now introduced, for example 1.85. Therefore, during the first comparison cycle, a high output of the comparator is not being interpreted as if $V_{in} > (1/2)V_{ref}$ but rather as if $V_{in} > (1/1.85)V_{ref}$. So the next comparison is with the middle of the next range $(447, \ldots, 1024)$, namely with code 735. Conversely, if the first comparison had resulted in a low output for the comparator, the next DAC code would instead have been 288. It can be seen that a redundancy (a scale extension) of $768 - 735 = 65$ codes $(288 - 256 = 65$ codes) or 65 LSBs has been introduced by changing the conversion base from 2 to 1.85.

The new non-binary bit weights $(447, 251, 142, 80, 45, \ldots)$ for the SAR operation are stored in a ROM and during each conversion cycle an ALU calculates the possible two codes for the next step. Depending on the comparator output, one of these is applied to the DAC.

The extended range has also other implications. First of all, since "less than one binary bit" is resolved for each cycle, more than n cycles are required in order to complete the conversion. For the case at hand, 12 cycles are required instead of 10. Second, an encoder transforming the 12 bits in base 1.85 into the desired 10 bits in base 2 is required.[6]

A different form of redundancy that can be introduced in order to address errors introduced by either comparator noise or limited ADC linearity issues has been described in [156]. An SAR ADC based on the charge-bi-section architecture previously discussed and shown in Fig. 3.14 is considered. As the previously explained bi-section process proceeds and the lower-order bits get resolved, the charge in the capacitors decreases and the DAC output presented to the comparator to determine the next bit becomes comparable to the RMS noise. Part of this noise is contributed by the dynamic comparator itself. Since the RMS input-referred noise of the comparator (as well as its input-referred

---

[6] Just like the digital error-correction logic in a pipelined ADC. This isn't typically a complex circuit.

offset) can be traded off against the comparator's power consumption [52, 33], a lower power (higher input-referred noise) comparator is being used while resolving the higher-order bits, while a higher-power (lower input-referred noise) comparator is used as the binary search converges toward resolving the lower-order bits [156]. This allows the comparator's power consumption to be reduced throughout most of the binary search, "shifting gear" to a more accurate/more power-hungry one as the search converges. Moreover, an additional comparison cycle is performed after the lowest LSB has been resolved. That allows one to detect whether a decision error had previously been made and, if that is the case, the LSB is inverted [156].

The redundant comparator is implemented by using two comparators, placed in parallel, equal in architecture but differently sized as mentioned above. They share the same inputs, but only one is enabled at each decision cycle and the corresponding outputs are read for its decision result.

Another source of power consumption that becomes non-negligible for very-high-speed sampling is associated with the generation and distribution of the clock used to time the internal operation of the SAR ADC. It has, in fact, previously been mentioned that, although the sampling clock f_s determines the rate at which the input is acquired, the SAR ADC needs to complete the bit-by-bit digitization cycles within $T_s = 1/f_s$ so that the next sample $\hat{V}_{in}(k)$ can be converted. So, if, for instance, $n + 1$ cycles are needed (one to sample the input and n to determine its n bits) then, internally, an $(n + 1) \cdot f_s$ clock is required to operate the SAR ADC. This can be onerous, in terms of design and power consumption, for a given technology when f_s is high enough. For instance, as remarked in [157], for $n = 6$ bits and $f_s = 300$ MSPS, internally a clock of $(n + 1) \cdot f_s = 2.1$ GHz needs to be generated and routed.

An alternate approach is to have the comparator itself trigger the next bit-conversion cycle as soon as the present bit decision has been taken. This results in an internal asynchronous operation that does away with the high-frequency clock generation and distribution (hence saving power), and speeds up the completion of the conversion itself since the next bit-conversion cycle can be started as soon as the present one has been completed (and, in fact, depending on the processed input, can even result in allowing a longer cycle time for some of the bits when the prior cycles have taken less than $1/((n + 1) \cdot f_s)$ seconds). This also stems from the observation that a dynamic comparator decides faster when its input is larger (i.e. in the SAR the DAC output and the reference have a large difference), as described by [157]

$$T_{cmp} = \frac{\tau}{A_0 - 1} \cdot \ln\left(\frac{V_{FS}}{V_{res}}\right) = K \cdot \ln\left(\frac{V_{FS}}{V_{res}}\right) \tag{3.3}$$

where A_0 is the small-signal gain of the internal inverting amplifier, τ is the time constant at the latch outputs, V_{FS} is the full logic swing level, and V_{res} is the voltage difference between the input signal and the reference level. Therefore, while a synchronous SAR ADC's conversion time is

$$T_{sync} = n \cdot K \cdot \ln\left(\frac{V_{FS}}{V_{min}}\right) \tag{3.4}$$

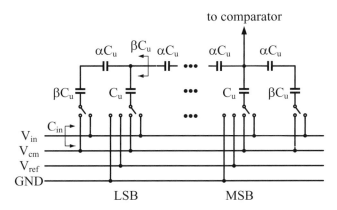

Figure 3.18 A non-binary radix capacitive DAC for the SAR ADC described in [157]. © 2006 IEEE.

where V_{min} is usually set by the LSB level (the smallest difference from the reference), the conversion time for asynchronous operation will be

$$T_{async} = \sum_{i=0}^{n-1} K \cdot \ln\left(\frac{V_{FS}}{V_{res}[i]}\right) \tag{3.5}$$

It has been shown in [157] that $T_{async}/T_{sync} \sim 1/2$, hence resolving bits asynchronously can be twice as fast as traditional synchronous conversion.

The SAR ADC described in [157] also uses redundancy in the form of a non-binary radix. However, in contrast with the previously discussed approach reported in [151], whereby the searched intervals were made to overlap in the digital domain, the approach described in [157] introduces the redundancy in the analog domain by means of a suitably ratioed capacitive network. More specifically, the non-binary (radix 1.81) capacitor array used both for sampling and for the DAC function is depicted in Fig. 3.18.

Instead of requiring only unit-sized capacitors C_u, this array requires three different sizes with ratios $1:\alpha:\beta$. The design equations for this network are

$$\beta = 1 + \alpha||\beta \tag{3.6}$$

$$\text{radix} = 1 + \frac{\beta}{\alpha} \tag{3.7}$$

The series capacitance connection dramatically reduces the effective input capacitance (e.g. $C_{in} \sim 90$ fF in [157]) of the network to

$$C_{in} = [1 + 2(\alpha||\beta)]C_u \tag{3.8}$$

The effect of the parasitic capacitance associated with the floating nodes can be reduced since α can be increased while maintaining the intended non-binary ratio. A foreground off-chip calibration scheme is used to determine the appropriate combination weights to compensate for the unpredicted parasitic capacitances and capacitor mismatch.

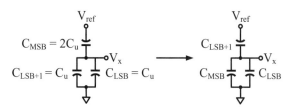

Figure 3.19 Down transition using the traditional charge-redistribution array [158]. © 2005 IEEE.

With this structure, a 6 b conversion requires seven comparison cycles (instead of six) and a 6 b 600 MS/s ADC was implemented in 0.13 μm CMOS consuming only 5.3 mW.

It was previously remarked that one of the major sources of power consumption in SAR ADCs is associated with charging/discharging the capacitors of the charge-redistribution network of Fig. 3.13. An important observation was made in [158], namely that the lumped binary weighted capacitors in the traditional charge-redistribution network do not make optimal use of energy to synthesize the desired DAC output. To illustrate that with a simple example, let us consider the case of a 2 b conversion. After the input V_{in} has been sampled (on all the bottom plates of the capacitors in the array), in order to resolve the MSB, the MSB capacitor C_{MSB} is tied to the reference voltage V_{ref} while all other capacitors in the array are tied to ground. That is shown on the left-hand side of Fig. 3.19.

Therefore, the resulting DAC output voltage presented to the comparator is

$$V_x = -V_{in} + \frac{1}{2}V_{ref} \tag{3.9}$$

During the next bit-cycle, one of two possible transitions is performed: if $V_x < 0$ then the input is greater than mid-scale and therefore the DAC output will need to increase, otherwise the DAC output will need to decrease. To accomplish the former (i.e. the DAC output needs to increase), C_{LSB+1} is switched from ground to V_{ref}, drawing

$$E_{up} = \frac{C_u V_{ref}^2}{4} \tag{3.10}$$

from V_{ref}. Conversely, if the DAC output needs to be decreased then C_{MSB} and C_{LSB+1} are switched as shown on the right-hand side of Fig. 3.19, and this results in drawing

$$E_{dn} = \frac{5 C_u V_{ref}^2}{4} = 5 E_{up} \tag{3.11}$$

from V_{ref}. Therefore, five times as much energy is used to perform the down transition. That's because all the charge previously supplied to C_{MSB} by the reference is dumped to ground, while C_{LSB+1} needs to be charged by the reference.

A better use of energy would result if, instead of dumping accumulated charge from some capacitors to draw additional charge into other capacitors, some of the existing charge were re-used to synthesize the next desired DAC output. To accomplish that, more freedom to use the capacitors is needed. So the binary weighted capacitor array is split into two sub-arrays by replacing the MSB capacitor C_{MSB} of the traditional network of Fig. 3.13 with a second copy of the remaining array as shown in Fig. 3.20.

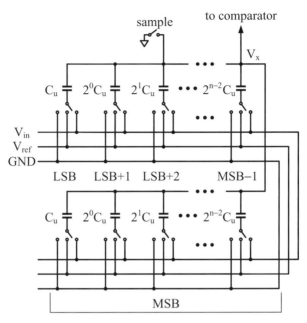

Figure 3.20 The "split-array" charge-redistribution DAC [158, 159]. © 2005 IEEE.

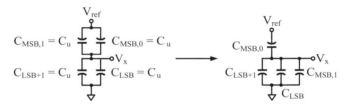

Figure 3.21 Down transition using the "split-cap" charge-redistribution array [159]. © 2005 IEEE.

Note that the overall capacitance of the new array (called a "split array") is the same as before, and so is the total area of the array.

To illustrate how the "split array" can be used to save energy the previous 2 b example is re-proposed. After the input has been sampled, during the first bit-cycle, just like before, the MSB capacitor (now made of $C_{MSB,0}$ and $C_{MSB,1}$) is tied to V_{ref} while all other capacitors are tied to ground, as shown on the left-hand side of Fig. 3.21.

So, the DAC output is unchanged and expressed by Eq. (3.9). If the input V_{in} is greater than mid-scale $V_{ref}/2$, then, just like before, C_{LSB+1} is switched from ground to V_{ref}, drawing the same energy E_{up} as in Eq. (3.10) from the reference. However, if the DAC output needs to be decreased, that is accomplished simply by discharging to ground half of the capacitors implementing C_{MSB} in the "split sub-array" as illustrated on the right-hand side of Fig. 3.21. Therefore, no new charge is being drawn from the reference and the energy involved this time is only

$$E_{dn} = \frac{C_u V_{ref}^2}{4} = E_{up} \tag{3.12}$$

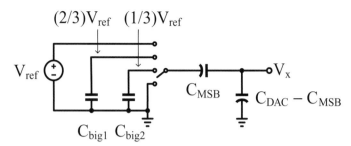

Figure 3.22 Step-wise charging of the MSB capacitor of the SAR's DAC [160]. © 2010 IEEE.

namely the same as if an upward transition had been required.

The extension of the operation to higher resolution and multiple cycles is straightforward.

The amount of energy saved by the "split-cap" approach depends on the nature of the signal converted and, as seen above, energy is saved (compared with the traditional approach) whenever a down transition is executed. Assuming a full-swing sinusoidal input distribution, the split capacitor array is expected to have 37% lower switching energy than the conventional array [159]. There is an additional advantage that is particularly valuable at high sampling rate: namely, with fewer capacitors being switched on down transitions, certain transient effects involving the array and the comparator do not occur, allowing faster settling and, correspondingly, some additional power savings in the comparator design [159].

An alternate way to save energy while charging and discharging the capacitive DAC has been introduced in [160].

Traditionally, to charge a capacitor C_{MSB} from zero charge (from "ground") to the reference voltage V_{ref}, the empty capacitor would be directly tied to the V_{ref} source. Similarly, to discharge it from V_{ref} to ground, it would be directly tied to ground. A full cycle from ground to V_{ref} and back to ground requires the source of V_{ref} to supply the following energy:

$$E_{DAC} = \frac{1}{2}C_{MSB}(0 - V_{ref})^2 + \frac{1}{2}C_{MSB}(V_{ref} - 0)^2 = C_{MSB}V_{ref}^2 \qquad (3.13)$$

However, if the charge/discharge process is done in S intermediate steps using other sources with potentials evenly distributed between V_{ref} and zero, then less energy is supplied by V_{ref} since the total supplied energy is distributed among all of the sources used.

An example of that is shown in Fig. 3.22, where the intermediate voltages $(2/3)V_{ref}$ and $(1/3)V_{ref}$ are provided by means of very large "charge-reservoir" capacitors C_{big1} and C_{big2}.

The example of Fig. 3.22 is analogous to what was described earlier for the "split-cap" array, namely C_{MSB} represents the MSB capacitor and its bottom plate needs to be charged, for example, from ground to V_{ref}, while the remainder of the DAC array, namely $C_{DAC} - C_{MSB}$, is grounded. Instead of tying C_{MSB} directly to V_{ref}, its bottom plate is tied first to C_{big2} (at $\sim(1/3)V_{\text{ref}}$), then to C_{big1} (at $\sim(2/3)V_{\text{ref}}$), and finally to V_{ref}.

When C_{MSB} is tied to C_{big2}, then, due to charge sharing, the voltage drops slightly below $(1/3)V_{\text{ref}}$. However, the drop is minor if $C_{\text{big2}} \gg C_{\text{MSB}}$. A similar observation can be made when C_{MSB} is tied to C_{big1}. However, the bottom plate of C_{MSB} is precisely equal to V_{ref} after it has been tied to that and fully settled.

Likewise, the discharging cycle of C_{MSB}'s bottom plate from V_{ref} down to ground is done in steps. C_{MSB}'s bottom plate starts at V_{ref}; then it is tied to C_{big1} and, because of charge sharing, the settled common voltage is slightly higher than the voltage on C_{big1} before being shorted to it (but not by a great deal if $C_{\text{big1}} \gg C_{\text{MSB}}$). Then, C_{MSB}'s bottom plate is tied to C_{big2} and drops one more step toward ground (and, again, because of charge sharing, C_{big2} experiences a slight increase of potential when the short is settled). Finally, C_{MSB}'s bottom plate is tied to ground and settled to zero accurately.

If the energy associated with the overhead is not accounted for, the energy used from the reference V_{ref} for this "step-wise" charge/discharge cycle with evenly spaced steps is only a fraction of the energy in Eq. (3.13) [161, 160]:

$$E_{\text{DAC, step-wise}} = \frac{C_{\text{MSB}} V_{\text{ref}}^2}{S} \tag{3.14}$$

Let us note the following [160].

- With regard to the DAC's accuracy, the accuracy of the voltages across C_{big1} and C_{big2} is not that critical. What is critical is the final value to which C_{MSB} settles after its charging cycle, namely that it accurately settles to V_{ref} and/or ground. Therefore, if, due to the charge-sharing process, these intermediate potentials vary a little as a function of the conversion, this is invisible in terms of the accuracy of the conversion process.
- Owing to the up and down cycling and to the charge-sharing processes, the two reservoir capacitors C_{big1} and C_{big2} pick up and drop some charge along the way. That charge is recycled to be used in the following charge/discharge cycles, and this is the essence of the energy saving of this approach.[7]
- Although these intermediate voltages vary during the operation, it can be seen that their limit (i.e. long-term) values are still actually the desired $(1/3)V_{\text{ref}}$ and $(2/3)V_{\text{ref}}$.

This approach, together with a few other techniques, was used to implement one of the most energy-efficient SAR ADCs published in recent years: at a sample rate of $f_s = 1 \text{ MS/s}$ and a supply voltage of 1.0 V, the 10-bit ADC prototype (ENOB = 8.85 b) implemented on a 65 nm CMOS process has been reported to consume 1.9 μW, achieving a value of Walden's figure of merit equal to FOM$_{\text{W}}$ = 4.4 fJ per conversion step [160].

---

[7] People familiar with hybrid cars, which store energy into a battery during braking and then retrieve it and use it later during regular driving or acceleration, may relate that to what is being discussed here.

Several other important techniques have recently been reported for SAR ADCs and, unfortunately, solely because their discussion would take up too much space, cannot possibly be included in this book.

As stated above, the SAR ADC architecture has been receiving a considerable amount of renewed attention due to its suitability for nanometer CMOS processes where decreasing supply voltages favor comparators (owing to the ability to achieve high gains, albeit strongly nonlinear with the input, and to use multiple cascade stages working in open loop) over linear amplifiers (as, for example, are necessary in a pipelined ADC).

There are also some non-negligible challenges that need to be considered when dealing with SAR ADCs. First of all, due to the many bit-decision iterations (at least $n + 1$; more if redundancy is introduced) required in order to complete the conversion the design and performance of internal clocking circuitry as well as the maximum speed of operation of the SAR can be quite severely limited for a given CMOS process node. Second, the large capacitive load represented by the S&H/charge redistribution DAC can become challenging to drive from a reference buffer and, perhaps more importantly, from the input signal source. In spite of the high power efficiency of the converter, the power spent in the internal clocking and the input/reference drivers can be larger than that expended by the converter itself.

SAR ADCs have also been extensively used as core architecture in several time-interleaved ADCs [162, 163]. The discussion of some of these is deferred to Section 3.4.

Research and product development centered on SAR ADCs continue at present to be among the most active areas in data conversion after a somewhat stagnant parenthesis during the nineties and early 2000s [164]. The dichotomous algorithm at the foundation of SAR ADCs has also been a source of inspiration of recent new converter ideas, such as the "binary search converter" [165] discussed in Section 3.3. Other notable recent examples of hybrids closely related to SAR ADCs include the SAR/pipelined ADC presented in [166] and the multi-bit SAR ADC described in [167].

Once again, this section aimed only to offer a partial snapshot of the dramatic recent evolution in this area of data conversion.

3.1.4   Continuous-time $\Delta\Sigma$ ADCs

$\Delta\Sigma$ converters are one of the most important converter architectures and among the oldest ones. Early $\Delta\Sigma$ modulators were, in fact, continuous-time systems [168, 169, 170, 171, 172, 173, 174]. Owing to technological advantages associated with the switch capacitor technique, discrete-time (switched cap) $\Delta\Sigma$ ADCs have been the dominant implementation for decades [81, 175]. However, recently, partly due to the repeatedly cited challenges associated with CMOS scaling and partly due to new developments and the increased popularity of calibration techniques (also enabled by the increased availability of digital functionality), continuous-time $\Delta\Sigma$ ADCs are rapidly coming back into play.

One of the aims of this section will be to discuss some of the differences between discrete-time (switched cap) $\Delta\Sigma$ ADCs (DT $\Delta\Sigma$) and continuous-time $\Delta\Sigma$ ADCs

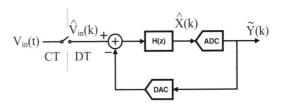

Figure 3.23 A discrete-time $\Delta\Sigma$ modulator.

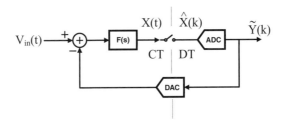

Figure 3.24 A continuous-time $\Delta\Sigma$ modulator.

(CT $\Delta\Sigma$). Our focus will be the modulator, namely the analog/mixed-signal part of the converter, rather than the digital back end performing the decimation and filtering.

Moreover, some peculiarities of CT $\Delta\Sigma$ ADCs and examples of their tuning will be briefly discussed.

There are important differences between DT and CT $\Delta\Sigma$ modulators. A DT $\Delta\Sigma$ modulator's block diagram is depicted in Fig. 3.23, while that of a CT $\Delta\Sigma$ modulator is shown in Fig. 3.24.

The switch shown in both diagrams symbolizes the place where the sampling and the transition from continuous time (CT) to discrete time (DT) happen. Consistently, continuous time is represented by the non-negative real variable t, while discrete time is represented by the non-negative integer variable k. Furthermore, a "hat" over a variable represents a sampled variable (e.g. $\hat{V}_{in}$), while a tilde over a variable represents a digital (quantized) variable (e.g. $\tilde{Y}$).

Many fundamental characteristics of $\Delta\Sigma$ ADCs are common and independent of the CT versus DT implementation details.

- Oversampling eases the problems associated with aliasing: since the bandwidth BW of the input signal $V_{in}(t)$ is much smaller than the Nyquist frequency $f_s/2$, the input signal aliases are greatly spaced in frequency from one another and therefore
  (1) the frequency attenuation roll-off requirements of an (analog) anti-aliasing filter placed in front of the sampler can be relaxed considerably; and
  (2) many of the distortion products haven't got sufficiently high frequency to fall out of the first Nyquist band ($f_s/2$) and therefore do not alias back onto the signal band (see also Fig. 3.25 suggesting also that, for example, the SFDR may be an ineffective metric with which to assess distortion in an oversampled converter).

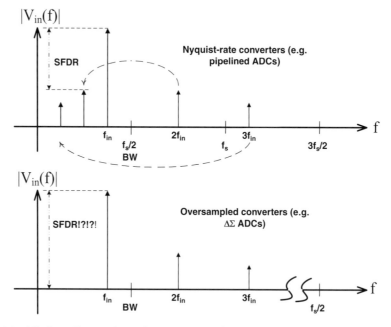

Figure 3.25 Aliasing of harmonics and measurement of SFDR in Nyquist-rate (top) versus oversampled ADCs (bottom).

- $\Delta\Sigma$ modulators allow much greater architectural flexibility than do traditional Nyquist ADCs, since they can be designed to have conversion characteristics that are lowpass, bandpass, quadrature etc. [81].

  In addition to that, CT $\Delta\Sigma$ ADCs have some more advantages.

- *Inherent anti-aliasing*: as depicted in Fig. 3.24 $X(t)$ has been filtered by $F(s)$ before being sampled at the input of the quantizer; this relaxes considerably the anti-aliasing filtering requirements on $V_{in}(t)$ [176], potentially completely eliminating it. This has extremely important consequences since anti-aliasing filters require board space and can require the introduction of active components, further adding noise and distortion and needing more power.[8]

- *Resistive input*: the input of a CT $\Delta\Sigma$ is often a resistor, as shown in Fig. 3.26; this is a fairly benign load for the circuitry driving the ADC and therefore it eases the input interface design. Conversely, switched capacitor input (Nyquist-rate or oversampled) circuitry first sinks a large impulsive current when the sampling switch is closed to the sampling capacitor and then kicks back the input signal source with charge injection when the switch opens. These signal-dependent impulses can introduce significant

---

[8] For example, in communication applications, it is common to use SAW or crystal filters. These filters are very linear and selective but introduce signal loss, which then needs to be compensated for by cascading linear, low-noise, and yet power-hungry amplifiers. Therefore, the inherent anti-aliasing property actually triggers a chain of "simplifications" in the processing chain ahead of the ADC that imply not only component elimination but overall system-level power savings.

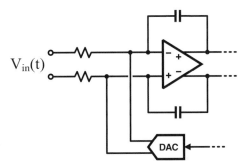

Figure 3.26 Typical input circuitry of a CT $\Delta\Sigma$ modulator.

distortion into the input signal before it is digitized and require driving circuitry that quickly recovers from the disturbances.

- *Suitability for low power*: the bandwidth of the active circuitry used in SC implementations (e.g. of $H(z)$ in Fig. 3.23) needs to be considerably higher than the processed signal frequency because of the need to accurately settle within the clock period [33]. That is not true in the CT modulator (particularly the loop filter $F(s)$ in Fig. 3.24), where the signals vary without significant sudden upsets (in other words, their bandwidth is lower) at the clock rate and can therefore be designed for lower power consumption.
- *Suitability for high bandwidth*: the previously cited advantage can be equivalently captured by noting that the bandwidth BW of the circuits used in a CT $\Delta\Sigma$ ADC can be closer to the f_T of the amplifiers (and devices) used in its modulator due to the considerably smoother waveforms being processed. Therefore, everything else being equal, CT $\Delta\Sigma$ ADCs can have higher BW than those of their DT counterparts [177].
- *Suitability for low-voltage and finer-lithography CMOS processes*: there are no switches in the loop filter $F(s)$. Moreover, unlike SC circuits requiring high-gain/high-headroom amplifiers with a well-behaved first/second-order closed-loop response, instead, the CT loop filter $F(s)$ can use multi-stage feed-forward amplifiers realizing high gain by means of a cascade of multiple low-voltage stages [35, 36, 37, 38, 39, 40, 178]. Implementations using inverter-based gain stages have even been disclosed [179]. This amenability to low-voltage supplies suggests that this architecture will not only fit better than others to CMOS scaling, but that, in fact, it will "thrive" by fully taking advantage of the higher f_T in these processes.

There are, expectably, also some drawbacks that need to be kept in mind as well. All $\Delta\Sigma$ (CT or DT) ADCs have the following disadvantages.

- *Digital post-filtering is required* since the output of the modulator needs to be decimated and the out-of-band noise needs to be digitally suppressed. With the increasingly lower cost of digital functionality, this disadvantage is rapidly becoming negligible.
- *Problematic overload condition*: all ADCs are driven in overload condition when their input goes above full scale, reacting by possibly giving several erroneous outputs and even taking some time to return to normal conversion behavior after the overload

condition has been removed. However, a $\Delta\Sigma$ ADC's overload condition occurs at an input level that is lower than the nominal full scale. Furthermore, for $\Delta\Sigma$ ADCs, recovering from an overload condition can be much more problematic than it is for other converters, especially if they become very unstable.

• *A very-high-frequency clock*: due to oversampling, the internal clock frequency can be very high. For example, an oversample ratio of 32 (OSR = 32) for a signal with a bandwidth of only 10 MHz (BW = 10 MHz) requires a sampling clock of $f_s = 2 \cdot$ OSR $\cdot$ BW $= 2 \cdot 32 \cdot 10 = 640$ MHz! The high-frequency sampling rate also implies that $\Delta\Sigma$ ADCs need very-high-frequency comparators, high-output-rate DACs, and high-speed digital logic (at least on the front-end part right after the quantizer, with this requirement being progressively relaxed as the decimation lowers the rate), in addition to high-speed analog circuitry in the modulator.

In addition, CT $\Delta\Sigma$ ADCs have also other disadvantages.

• *Increased sensitivity to sampling jitter*. It is well known that jitter in the sampling clock of an SHA causes a degradation of the noise floor of the sampled signal and hence of its signal-to-noise ratio [34, 52]. So, for example, for a sine wave $V_{in}(t) = A\sin(\omega t)$ sampled with a clock that has rms jitter (white and Gaussian) σ_j the corresponding SNR is

$$\text{SNR} = -20\log(\omega\sigma_j) \tag{3.15}$$

This improves thanks to oversampling (OSR) as[9] [81, 180]

$$\text{SNR} = -20\log(\omega\sigma_j) + 10\log(\text{OSR}) \tag{3.16}$$

In a CT $\Delta\Sigma$ the sampling jitter at the quantizer is negligible since its impact is reduced by the loop gain (thanks to the action of the feedback). However, it impacts the waveform at the output of the feedback DACs and, therefore, it is particularly sensitive for the DAC feeding back to the input of the modulator because it cannot be distinguished from $V_{in}(t)$. For a multi-bit DAC with M elements, and voltage output steps ΔV,[10] the jitter SNR with a -3 dBFS signal is [81]

$$\text{SNR} = 10\log\left(\frac{M^2 \cdot \text{OSR}}{4\sigma_{\Delta V}^2\sigma_j^2}\right) \tag{3.17}$$

This equation shows that the jitter noise degradation is always greater in a CT $\Delta\Sigma$ than in a DT $\Delta\Sigma$. On the other hand, this is mitigated by M (although the linearity of a multi-bit DAC can be an issue in itself), and it can be made comparable to a DT $\Delta\Sigma$. Furthermore, for high-frequency signals, when the signal changes by more than a few times the DAC LSB over a clock period, there is no difference between DT and

---

[9] It should be emphasized that this is the averaging result of oversampling and equally valid regardless of whether a $\Delta\Sigma$ ADC or a Nyquist-rate ADC is used to oversample the signal.

[10] The rms $\sigma_{\Delta V}$ of the output steps ΔV depends on the output waveform and the DAC quantization. For example, for a slow ramp, ΔV will correspond to the voltage granularity of the DAC, while for a rapidly varying signal the output will often skip intermediate quantization levels and jump rapidly with large ΔV steps.

CT in terms of jitter sensitivity [81]. This issue then tends to be more sensitive for lowpass $\Delta\Sigma$, but there is not much difference between DT and CT at high frequency, for example in the case of bandpass ADCs.

- *Tuning depends on f_s*: in a DT $\Delta\Sigma$ the transfer functions (signal transfer function, STF, and noise transfer function, NTF) scale with f_s, as in any switch capacitor filter. However, the STF and NTF in a CT $\Delta\Sigma$ depend only on the continuous-time filter $F(s)$, which is not clocked. Therefore, if f_s is varied then the *tuning of $F(s)$*, namely the action to control the position of the poles and zeros of $F(s)$, needs to be changed for the new value of f_s so that the newly scaled $F(s)$ matches the frequency response of the digital decimation filter suppressing the out-of-band quantization noise.

- *DAC switching behavior matters*: while in a DT $\Delta\Sigma$ what matters is only the DAC's output at the end of the clock cycle, conversely, in a CT $\Delta\Sigma$ the waveform is CT and therefore its entire shape, at any time, is crucial. Hence, the switching behavior of the DAC needs close attention since it can result in dynamic nonlinearity (Chapter 4 talks about this type of issue at length).

- *Excess loop delay*: another non-ideality of CT $\Delta\Sigma$ ADCs is the so-called excess loop delay (ELD), namely, referring to Fig. 3.24, the time that it takes to perform the sampling of $X(t)$, quantization ($\tilde{Y}(k)$), and digital-to-analog conversion ($Y(t)$). This non-zero time is due to the switching time in the comparators for the quantizer and in the DAC. Linearization techniques for the quantizer and the DAC, such as calibration and data-weighted averaging, can add even more delay to it. This delay needs to be accounted for in the implementation or it will detrimentally affect the transfer functions and the stability of the modulator. One of the main challenges is that while, on the one hand, it has a predictable fixed component that can be factored into the system design of the modulator, on the other hand, it is also a variable component that can be quite PVT-sensitive. Techniques to mitigate the impact of ELD do exist [181].

While CT $\Delta\Sigma$ ADCs are rapidly replacing their DT counterparts in low-frequency and narrowband applications such as for digital audio [182, 183, 184, 185] and in handsets [18, 186, 187], their introduction for higher-frequency applications, such as in medical applications [20, 178], TV [180, 188, 189], digital radio tuners [190, 191], and wireless communication infrastructure applications [177, 192, 193], is quite recent. That is partly due to recent advances in calibration techniques. Indeed, as mentioned above, the higher input signal frequency for such applications (of the order of tens of megahertz) implies that with today's most common processes (from 0.18 μm down to 45 nm) oversampling ratios (OSR) of the order of 16–32 are the highest practically possible (versus, for example, an OSR of the order of 256 as is typical for an audio ADC even in an older technology node). Therefore, to make up for the low OSR and in order to achieve relatively high dynamic ranges, the use of multi-bit DACs is necessary. Just like with traditional DT $\Delta\Sigma$ ADCs, which are assumed to be more familiar to the reader, using multi-bit DACs implies that even when only a limited number of output levels is synthesized, the linearity of the DAC (namely the accuracy with which each actual output level is close to its ideal output level) needs to be at least as good as the required ADC accuracy (that's the case for the DAC feeding back at the input of the modulator, while this requirement can be relaxed by an amount proportional to the intermediate gains between the input and other internal injection points for other possible DACs closing internal feedback loops). Similarly to what is discussed in Chapter 4 in the context of current-steering Nyquist-rate DACs, such linearity can be achieved by statistical design by means of intrinsic matching, or by calibration or using dynamic element matching (DEM). Typically, for signals of the order of tens of megahertz, while intrinsic matching is practical for total harmonic distortion of up to the mid 70s (dB), going beyond that will almost inevitably require calibration.

DEM is possibly another option. However, certain challenges need to be confronted if DEM is considered in this context. To begin with, the chosen DEM algorithm will need to be noise-shaped as well [189]. So, for example, a purely random scrambler [194] won't do the job since all the spurious tonal power will be simply distributed over the in-band noise floor, trading off THD for SNR but at no net improvement in SNDR. On using data-weighted averaging (DWA) [195, 196] or individual level averaging (ILA) [197] or other spectrally shaped algorithms, at least some of the spurious power is moved out of band and eventually filtered out in the digital domain. It should be pointed out, however, that these algorithms can also have some tonal behavior of their own and can require some additional randomization [198, 199]. Moreover, DEM algorithms, especially those which are noise-shaped, are truly effective when applied in conjunction with high OSR (which is the limitation we are battling to start with) and, most importantly, easily introduce additional latency (delay) due to the scrambler interposed between the DAC's input data and the actual DAC's output switches. Such delay might not be at all tolerable due to the above-mentioned issue with the excess loop delay and, therefore, low-latency implementations need to be considered [200].

Going back to calibration, unlike in DT $\Delta\Sigma$ ADCs, where the frequency placement of the poles and zeros of the modulator is essentially set by ratios of capacitors (hence, it is insured by matching/lithography), conversely, in CT $\Delta\Sigma$ ADCs poles and zeros are determined by ratios involving components of diverse nature (resistors, capacitors, inductors, and even active component parameters such as transconductances etc.) whose absolute value is then critical and that neither track each other over PVT variations nor can be made to have controlled mutual relations as a function of their geometry only. Therefore, in order to keep the performance and the response of the converter within specifications, it is necessary to control the position of the poles and zeros by means of specialized calibration, which is often referred to as *tuning*.

For example, let us consider an active-RC integrator as is typically found in lowpass CT $\Delta\Sigma$ ADCs, Fig. 3.27. As can be seen in this figure, the resistor is fixed while the variable capacitor is implemented as a digitally controllable binary weighted capacitor array:

$$C = C_0 + \sum_{k=0}^{6} 2^k \cdot C_u \cdot C\text{tune}[k] \qquad (3.18)$$

The RC time constant of these circuits needs to properly track the sampling clock f_s to make sure that the modulator's STF/NTF will track the digital filter's TF. The approach

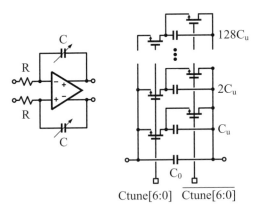

Figure 3.27 An example of an integrator with a programmable capacitor [178]. © 2006 IEEE.

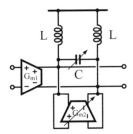

Figure 3.28 An example of a resonator with a programmable capacitor and a programmable Q-enhancement circuit.

described in [178] consists of finding the proper digital set point of the capacitor array by using a replica circuit and then using this code (or one with a known mutual relation) in the programmable capacitors for the integrators. The idea is as follows. A replica of the programmable capacitor C is fed with a constant current i for a time equal to a clock period $T = 1/f_s$, hence developing a potential of $i \cdot T/C$. This voltage is compared against a reference voltage $i \cdot R$, where R is a replica of the one used in the integrator and i is a copy of the current used for the capacitor C. Depending on the result of this comparison, C can be dialed up or down until, after iterations, the two voltages are made approximately equal, $i \cdot R \simeq i \cdot T/C$, and, hence, $RC \simeq T$.

Another example follows. An active-LC-tank resonator as is typically found in bandpass CT $\Delta\Sigma$ ADCs is depicted in Fig. 3.28. Ideally its resonance frequency is set by the two inductors L and the programmable capacitor C while the negative transconductance block $-G_{m2}$ (a cross-coupled differential pair) compensates for part of the LC-tank loss in order to enhance the quality factor Q of the resonator [201]. Let us say that, for example, its resonance frequency needs to be tuned so that, for example, it is centered at $f_s/8$ [188]. To achieve that, the input to the driving transconductor G_{m1} is nulled while $-G_{m2}$ is increased until the LC tank begins oscillating. The frequency of oscillation is compared against $f_s/8$ and the programmable capacitor (which can be implemented in a similar fashion to the previous example) is varied accordingly, for example, through a binary

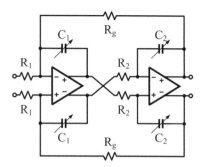

Figure 3.29 An example of an active-*RC* resonator with programmable capacitors.

search algorithm (similarly to what is done in an SAR ADC). A conceptually simple way to compare the frequencies is to use a digital counter to count how many periods of the waveform at the output of the oscillating resonator are present over a prescribed number of clock periods [189] or, alternatively, to use a frequency detector similar to one of those used in PLLs. Since the frequency drift of the *LC* tank is minimal over temperature, it is often sufficient to perform the calibration/tuning only at power-up [188].

Alternatively, a resonator can also be implemented by cross-coupling two active-*RC* tuned integrators (see Fig. 3.29), like those already discussed in the first example, as reported in [180] and [189].

Multi-bit DAC linearity and clock jitter sensitivity, loop filter linearity, noise performance, and ability to tune the transfer function over a large clock range are non-negligible challenges that have only recently been receiving renewed attention. Conversely, switch-capacitor-based converters (Nyquist-rate and oversampled) benefit from decades of experience and development. Nevertheless, thanks to their compelling advantages in terms of system-level simplification (elimination or simplification of the front-end circuitry and filtering) and their fitness for nanometer-scale CMOS processes, CT $\Delta\Sigma$ ADCs are beginning to rival established architectures such as pipelined ADCs in their prime application and performance space (instrumentation, industrial, wireless, and medical applications among others). It is not unthinkable that, as converter as well as process technology progresses, and as some of these technical challenges are met by new techniques, we may witness a "migration" of pipelined ADCs toward a higher-signal-bandwidth and, perhaps, lower-dynamic-range space, while CT $\Delta\Sigma$ ADCs increasingly move over into the performance space at present dominated by pipelined ADCs. More importantly, the ability of $\Delta\Sigma$ ADCs to perform special types of conversion, such as bandpass [187, 188, 202, 201] and complex [18, 191] conversion, has the potential to enable new signal processing structures and/or new applications that haven't yet emerged.

3.1.5   Incremental and extended counting ADCs

Incremental ADCs [6], also known as "one-shot" [203], "no-latency," or "charge-balancing" $\Delta\Sigma$ ADCs, can be thought of as a special class of $\Delta\Sigma$ ADCs. Although

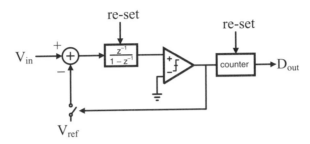

Figure 3.30 A first-order incremental ADC. © 2004 IEEE.

they are structurally very similar, the main difference between incremental ADCs and traditional $\Delta\Sigma$ ADCs is that, while $\Delta\Sigma$ ADCs operate continuously, processing time-varying signals,[11] incremental ADCs instead process each sample at one time and both the analog and the digital memory elements (e.g. integrators in the analog modulator and counters in the digital filter) of an incremental ADC are re-set before each new conversion cycle. Furthermore, in an incremental converter the decimating filter following the $\Delta\Sigma$ modulator can be implemented with much simpler structures than in traditional $\Delta\Sigma$ ADCs. For example, that can be just a counter in the case of a first-order incremental ADC. In fact, it can be shown [6] that a first-order incremental ADC, like the one depicted in Fig. 3.30, is nothing other than a classic dual-slope converter where the integration of the input and the reference are performed alternately instead of separately. The voltage at the output of the integrator at the end of the conversion cycle is an analog representation of the conversion residue (namely, the continuous input voltage minus its quantized representation, $V_{in} - \tilde{V}_{in}$).

Although there are not many publications in the open literature, incremental ADCs actually find wide commercial use in a broad range of instrumentation and measurement applications (e.g. digital voltmeters as well as a number of sensor applications [204]) where *absolute* accuracy is required (i.e. offset and gain errors cannot be tolerated) [6]. While the accuracy of these converters can be quite high (often in excess of 16 ENOB), particularly if they use a single-bit feedback DAC, their sample rate is rather low (often lower than tens of KSPS) and so is their power consumption and area. Incremental ADCs are part of what is commonly referred to as "precision converters," and just like with traditional $\Delta\Sigma$ ADCs, it is possible to have higher-order loops, multi-bit architectures, cascaded/MASH architectures etc. with the corresponding advantages and trade-offs in accuracy, speed, power, complexity etc.

Incremental ADCs have also been finding attention in the form of hybrid architectures, for example, in [205] (an incremental–SAR hybrid), [206] (an incremental–algorithmic hybrid), and [207] (an incremental–algorithmic hybrid) "extended counting ADCs" are described. In [207] the proposed micro-power extended counting ADC is configured to work as an incremental ADC during the first phase of conversion ("counting

---

[11] We could call them "convolutional" ADCs since they process an entire waveform, instead of individual, possibly uncorrelated, samples, and, therefore, keep some memory of prior input values over time.

conversion"), obtaining the first residue of the conversion, and then completes the conversion by reconfiguring the same hardware to work as an algorithmic converter ("extended conversion") further quantizing the prior residue to quickly resolve the remaining LSBs. That is done to trade off the high accuracy (with slow conversion time) of an incremental ADC with the higher speed (but lower accuracy) of an algorithmic ADC. Owing to its compactness, accuracy, and low power consumption, an extended counting ADC was recently used as a key block in a CMOS single-chip electronic compass sensor including also the Hall sensors and the complete digital signal processing for accurate heading calculation [208].

3.2   Emerging architectures and techniques

Using new techniques to dust off and resurrect older architectures has not been the only avenue that ADC designers have been taking in order to deal with the challenges of CMOS scaling and increasing performance demands. Indeed, other profoundly different architectures and techniques have also been proposed recently. Although, inevitably, some of the fundamental principles and ideas of data conversion are still present as the soul of some of these new schemes, their embodiment is certainly breaking away from the classic schemes discussed earlier.

There are indeed common goals that lead researchers and engineers to look into introducing and developing the architectures discussed in the following. These can be identified with the need to address two main, previously discussed, challenges:

- consequences of CMOS technology scaling, hence, primarily, lowering supply voltages/headroom, lowering a transistor's intrinsic gain (g_m/g_{ds}), increasing the transit frequency (f_T), and increasing gate and subthreshold leakage currents;
- the quest for ever higher conversion efficiency, namely lower energy and/or lower power consumption, for a given set of dynamic performance specifications.

As pointed out in Section 1.2, these two driving forces are sufficiently fundamental in their impact as to send engineers back to the proverbial "white board" to figure out what had to be re-thought or replaced (the architectures, the building blocks, the devices?) in order to adapt and thrive on this technology road-map.

Some examples of promising emerging solutions are discussed in this Section.

The zero-crossing technique covered in Section 3.2.1 tries to provide an answer to the question "can we continue designing switch capacitor circuits without using operational amplifiers?"

The time-to-digital (TDC) converters (and digital-to-time converters) of Section 3.2.3 have actually been around, more or less explicitly or latently, for longer than most might think. Fundamentally, given that headroom is narrowing down but, potentially, device speed is rapidly increasing, these try to address the question "can our analog processing variable be time, as opposed to voltage or current?"

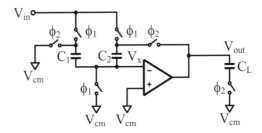

Figure 3.31 A traditional switch capacitor op-amp-based gain circuit. © 2007 IEEE.

The philosophy behind digitally assisted analog techniques of Section 3.2.4 can be summarized as follows: since digital functionality abounds, let us leverage it to compensate for the analog challenges from the digital domain.

Lots is happening, and very rapidly, in all of these breakthrough areas. As could be expected, there are probably more questions than answers. In fact, many important questions might not yet have been asked. So, please be advised that the opportunities are significant but this is uncharted territory, you're entering at your own risk!

3.2.1   Comparator-based/zero-crossing ADCs

As pointed out several times in this book, as CMOS technology scales, gate oxide thins and supply voltages drop, a fundamental building block of most ADCs and some of the voltage-domain DACs becomes dramatically hard to design: the operational amplifier. Op-amps enable robust, accurate, linear, low-noise circuits when used in negative-feedback configurations. That allows accurate, PVT-insensitive transfer characteristics that primarily depend on the matching properties of passive components or geometrically ratioed active components.

Feedback circuits completely rely on op-amps to provide very high open-loop gain and stability in closed loop. Unfortunately, these two elements often conflict with each other and become extremely difficult to conjugate in deep nanometer CMOS processes [209].

A traditional switch capacitor gain circuit based on using an op-amp in negative feedback is depicted in Fig. 3.31 [33].

On ϕ_1 the input V_{in} is sampled on C_1 and C_2, while the op-amp inputs are both shorted to the differential ground V_{cm}. On ϕ_2, capacitor C_2 is closed in the negative feedback path of the op-amp. The feedback action drives V_{out}, and this in turn, by means of the capacitive feedback network constituted by C_2, drives the amplifier's negative input voltage V_x toward the positive input held at V_{cm}. Since the top plate of C_1 is shorted to V_{cm} and its bottom plate is driven to V_{cm} by the feedback action, the charge held in C_1 is forced to transfer into C_2 and the output voltage V_{out} develops:

$$V_{out} = \left(1 + \frac{C_1}{C_2}\right) V_{in} \qquad (3.19)$$

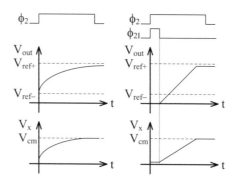

Figure 3.32 Typical waveforms for traditional op-amp-based (left-hand side) and comparator-based (right-hand side) switched capacitor gain stages. © 2007 IEEE.

The latter is sampled on C_L when its series switch opens at the end of ϕ_2. The accuracy with which Eq. (3.19) represents a real implementation depends on the actual value of V_x at the end of ϕ_2 (the steady-state error is inversely proportional to the open-loop DC gain of the op-amp A_{dc} while its settling depends primarily on the initial slew rating and the closed-loop time constants during the following linear exponential settling phase). Examples of waveforms are depicted on the left-hand side of Fig. 3.32.

What is really important, for this circuit, is that an accurate V_{out} is realized at the end of ϕ_2 when that is sampled on C_L. The gist of the op-amp function is to drive V_x toward V_{cm}, and the resulting V_{out} is the effect of that.

However, the very same final result would be obtained if V_x were somehow forced to *cross* V_{cm} and the circuit stopped varying V_{out} as soon as this condition (that is $V_x = V_{cm}$) had been *detected*.

This is the fundamental idea behind a new and emerging class of switch capacitor circuits that operate *without* op-amps, and it relies on being able to *detect* the *crossing* of a virtual ground (a "*zero* voltage").

The first implementation of this idea was the so-called *comparator-based* switched capacitor (CBSC) technique [210], whereby the detection of the zero-crossing (i.e. the detection that V_x has reached V_{cm} and, therefore, V_{out} has reached the desired final value) is performed by a comparator.

The switched capacitor gain stage of Fig. 3.31 becomes the CBSC gain stage of Fig. 3.33, where the op-amp has been replaced by a comparator and a controlled current source. For ease of comparison, simplified waveforms for this circuit are depicted on the right-hand side of Fig. 3.32, next to those of the traditional op-amp-based gain stage.

While the sampling phase during ϕ_1 is analogous, this circuit differs from the op-amp-based circuit during ϕ_2. Specifically, at the beginning of ϕ_2 the switch S_2 is briefly closed (on ϕ_{21}) precharging V_{out} to the AC ground V_{cm}. Also, because of this switch's action on C_2's top plate as well as the closing of the switch tied to the top plate of C_1, V_x is pushed down to $V_{cm} - V_{in}$. Once S_2 is released to open, the constant current source I_1 begins charging C_2's top plate, and hence both V_{out} and V_x (via the capacitor divider constituted by C_1 and C_2) linearly ramp up. Once V_x reaches V_{cm} the comparator detects the virtual

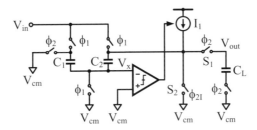

Figure 3.33 A comparator-based switch capacitor gain circuit [210]. © 2007 IEEE.

ground condition and stops the current source I_1, completing the charge transfer from C_1 to C_2.[12]

It is important to remark again that, although the waveforms in Fig. 3.32 are very different, what really matters is not how V_{out} evolves toward its final value, but how accurate its final value is. This depends on the accuracy of the virtual ground condition at V_x when the sampling switch opens and V_{out} is sampled on C_L. In the circuit of Fig. 3.33 the comparator only *detects* the zero-crossing condition, it does not force the virtual ground condition. Therefore it is much more power efficient than the circuit of Fig. 3.31 using an op-amp as well as more amenable to low supply voltage since the DC gain–stability trade-off of op-amps is resolved with comparators.

The next evolutionary step to the CBSC technique was the *zero-crossing*-based (ZCB) technique [211]. That stems from the observation that, in general, voltage comparators are designed to compare two arbitrary voltage waveforms, while in this technique the aim is simply to detect the zero-crossing condition where one of the two inputs is a (uni-directional) linear ramp. This specification allows one to replace the comparator with a simpler and more power-efficient *zero-crossing detector* (ZCD).

This next evolution is shown in the ZCB gain stage of Fig. 3.34.

The ZCB circuit of Fig. 3.34 differs from the CBSC circuit of Fig. 3.33 simply by the replacement of the comparator with the dynamic inverter composed by transistors M_1 and M_2. Its behavior is analogous to that of the CBSC gain stage. On ϕ_1 the input V_{in} is sampled on C_1 and C_2. At the beginning of ϕ_2, during ϕ_{21}, M_4 precharges V_{out} to ground. Again, V_x is pushed down to $V_{cm} - V_{in}$ and hence M_1 turns off. At the same time M_2 gets turned on, lifting V_p to V_{dd} and then turning on M_3. C_L is therefore initialized below full scale. Once the re-set phase ϕ_{21} is completed, V_p remains charged high and keeps the sampling switch M_3 on, while V_{out} begins to ramp up due to the current I_1. Once again, V_x ramps up as well according to the capacitor divider constituted by C_1 and C_2. At some point V_x becomes high enough that M_1 begins conducting current and V_p drops down. M_3 turns off at this point, the output sample is taken, and the charge transfer is completed.

---

[12] The first CBSC gain stage published in [210] was actually slightly more complex. It had a *coarse* charging phase during which I_1 was large, to drive V_x close to V_{cm} very rapidly, followed by a *fine* charging phase during which the ramp approached the virtual ground condition slowly, hence facilitating accurate detection of the crossing condition.

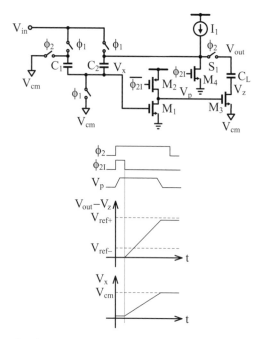

Figure 3.34 A zero-crossing-based switch capacitor gain circuit and associated waveforms [211].
© 2007 IEEE.

Note that, since the slope of the V_x ramp created by I_1 is constant, M_1 consistently switches always at the same voltage. This threshold is PVT- as well as ramp-dependent, but it is not input-signal-dependent. So it introduces an offset (which can be corrected in a number of ways) but not distortion. The main drawback of this circuit is its being single-ended and hence suffering from all the usual limitations in terms of linearity, noise, and disturbance rejection, and hence it is rather unsuited to high-resolution applications. However, it is certainly much more amenable to low-supply operation and considerably more power efficient and faster than the CBSC approach [211].

A 1.5 bit/stage pipelined ADC was implemented with multiplying switched capacitor DACs (MDACs) using the ZCB technique in a 0.18 μm CMOS technology. This performed with 6.9 b ENOB (6.4 b) at 100 MSPS (200 MSPS) and Walden's figure of merit was equal to $FOM_W = 0.38$ pJ per conversion step (0.51 pJ per conversion step) [211].

A fully differential ZCB 10 b 26 MSPS pipelined ADC was subsequently demonstrated in 65 nm CMOS [212]. The fully differential version had two current sources to implement the coarse and fine charge as in the first CBSC pipeline. More importantly, the extra power spent in the differential ZCD (implemented as a differential pre-amplifier driving a dynamic threshold detector) allowed improvement of the dynamic performance offered by differential circuits. This performed with 8.73 b ENOB at 1.2 V supply and Walden's figure of merit was equal to $FOM_W = 0.16$ pJ per conversion step [212].

Other noteworthy ZCB ADCs have recently been published, including a 1.1 V ZCB ΔΣ ADC using switched resistor current sources in 45 nm CMOS, with an input bandwidth

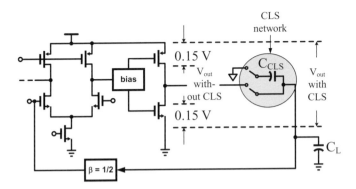

Figure 3.35 A low-voltage, low-gain amplifier used in an SC circuit. Without using CLS the output swing of the amplifier is limited by the need to keep the output transistors in saturation ($V_{ds} > V_{dssat}$). The CLS capacitor C_{CLS} introduces an output-dependent level shift into the voltage on the load capacitor C_L and extends the corresponding swing on it [217]. © 2008 IEEE.

of 0.833 MHz, 7.63 b ENOB and $FOM_W = 1.91$ pJ per conversion step [213], and a 1.2 V 12 b 50 MSPS fully differential pipelined ADC in 90 nm CMOS with 10 b ENOB and $FOM_W = 88$ fJ per conversion step [214].

The rapid improvement in performance, especially in terms of power efficiency, is visible and can be expected to continue for some time as this techniques matures and some of the practical shortcomings become better understood and addressed [215, 216]. Theoretical projections made in [209] indicate the potential for ZCB ADCs to be 3–10 times more power efficient than traditional ADC architectures with the same SNDR and sample rate.

So far, however, due to various practical limitations (e.g. ramp linearity limiting the higher resolution, PVT sensitivity of various sorts, etc.) only prototypes with ENOB between ~7 b and ~10 b and sample rates in the tens of MSPS have been demonstrated in the open literature.

3.2.2 Correlated level-shifter ADCs

Another recently introduced technique dealing with the voltage headroom restrictions in deep nanometer CMOS processes is the so-called "correlated level shifter" (CLS) technique [217]. This technique is used in switched capacitor (SC) circuits to increase both the effective voltage range and the effective DC open-loop gain of the amplifier. Particularly because of the latter it has some similarities to the well-known correlated double sampling (CDS) [218]. An example of such a circumstance is shown in Fig. 3.35.

The voltage swing on the output stage is bounded by the need for both output transistors to stay saturated ($V_{ds} > V_{dssat}$). Failing to achieve that causes significant (nonlinear) loss in the amplifier gain and output compression. Moreover, due to the headroom limitations (consider a supply of the order of ~1 V or less) cascoding and gain boosting are not possible and therefore the amplifier has a low DC voltage gain (say ~30 dB). These issues are addressed in the CLS circuit example shown in Fig. 3.36.

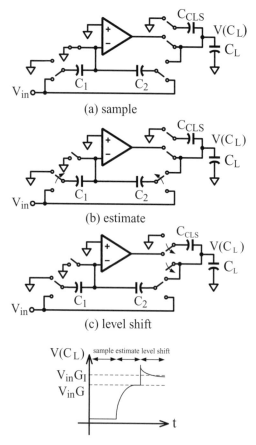

Figure 3.36 An example of a switch capacitor amplifier with desired gain of 2 using the CLS technique [217]. © 2008 IEEE.

The circuit operates in three phases called "sample," "estimate," and "level shift." During "sample" the input voltage is sampled in $C_1//C_2$ while C_{CLS} and C_L are discharged to ground. During "estimate" the circuit behaves as a traditional SC amplifier with finite op-amp gain A_{EST} and a closed-loop voltage gain G of

$$G = \left(1 + \frac{C_1}{C_2}\right)\left(\frac{1}{1+1/T}\right) \qquad (3.20)$$

where $T = A_{EST}C_2/(C_1 + C_2 + C_{IN})$ while $G_I = (1 + C_1/C_2)$ is the ideal closed-loop gain one would want and that would result from an infinite DC gain A_{EST}. But since the DC gain A_{EST} is small for the above-mentioned reasons (e.g. $A_{EST} \sim 30\,\mathrm{dB}$), the output voltage on the load transistor C_L (and on C_{CLS}) settles toward a final value with a steady-state error that is inversely proportional to the op-amp DC gain A_{EST}. Hence the output, in this case, is an "estimate" of the actual desired output. Finally, during the "level shift" phase the CLS capacitor C_{CLS}, which is holding the estimated output, is reversed and interposed between the op-amp output and the load capacitor. On doing this the node voltage at the load capacitor C_L first jumps up as a result of the capacitive level shift introduced by C_{CLS} and then begins settling down toward a new steady-state value *closer* to the corresponding ideal gain $G_1 = (1 + C_1/C_2)$. It can be proved, in fact, that the effective gain due to the CLS is [217]

$$G_{CLS} = \left(1 + \frac{C_1}{C_2}\right)\left(\frac{1}{1 + 1/T_{EQ}}\right) \tag{3.21}$$

with

$$T_{EQ} = T(2 + T) \sim T^2 \tag{3.22}$$

This is equivalent to having an op-amp with DC gain squared up (e.g. $A_{EQ} \sim 60\,\mathrm{dB}$ in the previous example). Moreover, the voltage at the capacitive load C_L, thanks to the level shift, can now actually rise or drop beyond the limits introduced by the operational amplifier and can reach (or even go beyond) the supply voltage. This is a non-negligible advantage since with a higher signal swing it is possible to achieve an equal signal-to-$k_B T/C$ ratio (thermal SNR) with a much lower capacitance. Hence the circuit can be made to work fast.

A 12 b 20 MSPS pipelined ADC using CLS for the implementation of the MDACs/pipelined stages has been presented in [217] as a proof of concept in which, on turning the CLS on and off, it is possible to see a clear improvement in linearity due to the gain-magnifying effect and range-magnifying effect of the CLS.

3.2.3   Time-to-digital and digital-to-time converters

Another class of mixed-signal circuits that has recently seen renewed attention due to its potential advantages in nanometer CMOS technologies is the class of time-to-digital (TDCs) and digital-to-time (DTCs) converters. These are analogs of ADCs and DACs, respectively, but, instead of having voltages or currents as the analog variables, in TDCs and DTCs the analog variable is time, or, more precisely, time intervals. As stated, this type of circuit has actually been around for decades, in various incarnations, as building blocks in instrumentation and communication circuitry [220, 221, 222, 223] and in "all-digital" PLLs and DLLs [224]. So, perhaps, this subsection should have been placed in Section 3.1. However, in the past, these architectures had not been fully recognized as data converters and even nowadays there is still some debate on this issue. Because of that, perhaps, this is in fact the appropriate place for this topic.

TDCs and DTCs are usually almost entirely constituted by asynchronous digital blocks and by analog timing circuitry that does not need to accurately process continuously varying voltages/currents but rather deals with the edges of square-wave-like waveforms or rectangular pulses. Combining that with the observation that with CMOS scaling the transition delays of logic gates decrease and it is then easier to generate and accurately process short-time pulses, it is conceivable to conclude not only that these semi-digital circuits scale well but also that the ability to discriminate time intervals with increasingly higher resolution (and potentially lower phase noise) improves on moving to finer and finer lithography nodes. As pointed out above, these are very interesting potentials and

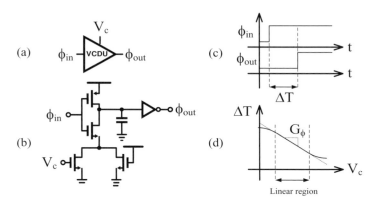

Figure 3.37 (a) A voltage-controlled delay unit (VCDU); (b) a possible circuit implementation based on a current-starved inverter; (c) the time diagram of the input ϕ_{in} and output ϕ_{out}; and (d) an example of the voltage-to-phase characteristic of a VCDU, where a central linear region can be identified with a corresponding linearized voltage-to-phase gain G_ϕ [223]. © 2010 IEEE.

have generated quite some attention and research activity, leading to notable advances in a relatively short time. It is, however, important to remark that, due to practical limitations being progressively addressed, the potentials have yet to be fully realized and the resolution and bandwidth of TDCs and DTCs are still not very competitive with many state-of-the-art ADCs and DACs.

This section will provide a brief introduction to TDCs and DTCs, with some recent examples of such converters.

The simplest and most common blocks used in TDCs and DTCs are the voltage-controlled delay unit[13] (VCDU) and the D-type edge-triggered flip-flop (DFF). A VCDU is any circuit able to take a rectangular pulse as an input and provide the same rectangular pulse as an output with a time delay that can be controlled (within a certain range) by a control voltage V_c as depicted in Fig. 3.37.

VCDUs are commonly found in ring-oscillators (e.g. in PLLs) and voltage-controlled delay lines (e.g. in DLLs) and can also be used to convert input voltages into time intervals since, inevitably, many signal sources are in the form of voltages or currents. The DFF can be used as an analog for a comparator since, given two pulses, say ϕ_{in} and ϕ_{ref}, fed to its D input and clock input, respectively, as shown in Fig. 3.38, it will return a logic 1 at its Q output when ϕ_{in} leads ϕ_{ref} ($\phi_{in} < \phi_{ref}$) and 0 otherwise ($\phi_{in} \geq \phi_{ref}$).

Using these simple blocks it is possible to build a variety of DTCs and TDCs.

For example, a simple DTC can be built using a DLL as depicted in Fig. 3.39 [154, 223]. In this 3 b DTC example the DLL generates a set of evenly spaced square waves (phases) B_0–B_7 locked to the reference clock ϕ_{ref} ($B_0 = \phi_{ref}$) depicted in the time diagram below the circuit scheme; these phases are then fed to a multiplexer (or another suitable digitally controlled combiner if a different pulse shape is desired), which selects

---

[13] Indeed, there are also current-controlled delay units and most of what will be stated below can be equivalently re-proposed with currents instead of voltages.

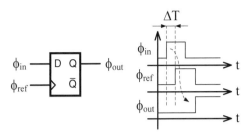

Figure 3.38 An example D-type positive edge-triggered flip-flop (DFF) used as a time-mode comparator.

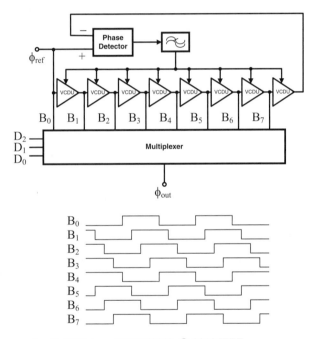

Figure 3.39 An example of a DLL-based DTC [223]. © 2010 IEEE.

the desired time-delayed pulse that is routed to the output ϕ_{out} using the input digital code $D_{\text{in}} = [D_0 D_1 D_2]$.

Considering TDCs, starting from the simplest converter architecture, a time-mode flash converter is shown in Fig. 3.40 [222, 223]. In this 3 b time-mode flash converter, the comparators have been implemented using DFFs while the reference resistor ladder has been implemented by cascading identical VCDUs with the same control voltage V_{c} and the same individual delay τ. The input ϕ_{in} is fed to all the DFFs and depicted in the top left corner of the figure. The reference ϕ_{ref} is fed to the first comparator and then progressively delayed through the voltage-controlled delay line where it is tapped to each of the remaining DFFs. The corresponding waveforms are represented on the

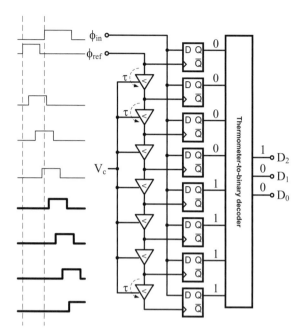

Figure 3.40 A flash time-to-digital converter [223]. © 2010 IEEE.

left of each VCDU. In the example shown in Fig. 3.40 the time difference between the positive edge of ϕ_{in} and ϕ_{ref} falls between 3τ and 4τ. As a result, the top four DFFs give a zero output while the bottom four DFFs give a one output. The corresponding thermometric output of the DFFs (accurate timing is omitted in this figure for simplicity of explanation) is then provided to a thermometer-to-binary decoder, which returns the 3 b output D_{out}.

In order to better control the time-delay of the VCDUs for PVT variations and to insure that, aside from cell-to-cell mismatches, the period of ϕ_{ref} is evenly divided by the delays τ, it is possible to complete the delay line with a phase detector and a filter closing it into a delay-locked loop (DLL) as shown in Fig. 3.41 [222, 223].

One of the limitations of the previous architecture is that the time resolution is limited by τ (which, being a gate delay, is basically limited by the speed of the process technology). One approach allowing discrimination of smaller time intervals is to use a so-called "Vernier" delay line [225] as shown in Fig. 3.42. It can be seen that a second identical delay line is introduced and the D inputs of the DFFs are tapped from its VCDUs. The DFFs now see increasing multiples of the differences of the delays of the two chains:

- the top DFF sees ϕ_{in} and ϕ_{ref};
- the next one sees ϕ_{in} delayed by $\tau_2(V_{c2})$ and ϕ_{ref} delayed by $\tau_1(V_{c1})$, hence that is equivalent to having delayed ϕ_{ref} by a net $\tau_1 - \tau_2$ seconds, bringing it closer to ϕ_{in} by that amount;

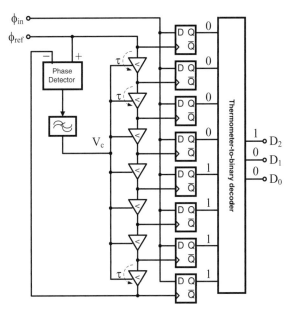

Figure 3.41 A flash time-to-digital converter with the time reference ladder closed in a DLL [223]. © 2010 IEEE.

- the following one sees ϕ_{in} delayed by $2\tau_2(V_{c2})$ and ϕ_{ref} delayed by $2\tau_1(V_{c1})$, hence that is equivalent to having delayed ϕ_{ref} by a net $2(\tau_1 - \tau_2)$ seconds;
- and so on.

The key point is that the progressive delay can now be controlled to a finer granularity $\tau_1 - \tau_2$. It is implicit that V_{c1} and V_{c2} need to be controlled by DLLs as has previously been shown. Clearly, mismatches do limit the approach.

Again, continuing with analogies from known converters' architectures and techniques, it is possible to increase the resolution of the delay line by means of phase interpolation, similarly to what is done when performing interpolation in folding and interpolating ADCs or in flash ADCs. An example of phase interpolation is depicted in the scheme shown in Fig. 3.43 [226, 227]. This technique can, for example, equally well be applied to the flash TDC of Fig. 3.40 (e.g. reducing the number of VCDUs but not the number of DFFs for a constant number of bits or, alternatively, increasing the resolution by increasing the number of DFFs with a constant number of VCDUs) and also to the DTC of Fig. 3.39 (e.g. increasing the number of phases fed to the multiplexer by interpolating between those generated by the DLL; the multiplexer, in this case, will be controlled by a higher number of bits).

As expected, the corresponding limitations are going to be analogous to those in voltage-mode interpolators.

Let us consider another example of TDC. It is well known that frequency is the derivative of phase ($f = d\varphi/dt$). So, for example, the phase difference between two

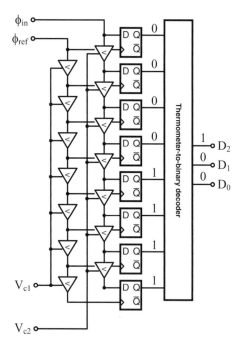

Figure 3.42 A flash time-to-digital converter with finer time resolution using a "Vernier" delay line [223]. © 2010 IEEE.

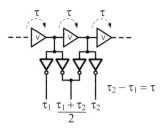

Figure 3.43 An example of phase interpolation. Adapted from [227].

oscillators with a fixed (controlled) frequency difference $f_1 - f_2$ will drift in phase at a constant rate. Likewise, a voltage-controlled oscillator (VCO) can be used as a time-mode voltage-controlled phase integrator and therefore is a suitable building block for a TDC requiring integrators such as a $\Delta\Sigma$ TDC. An example of a first-order $\Delta\Sigma$ TDC and associated waveforms is shown in Fig. 3.44 [220, 223, 228, 229, 230] ($\varphi(\phi_o)$ represents the phase of the VCO's output ϕ_o in radians). The TDC is constituted by a traditional voltage-mode sample-and-hold (SHA) whose sampled output $V_c(n)$ drives a VCDU-based ring-oscillator VCO[14] with output waveform ϕ_o. The latter is fed to a

---

[14] Please pay attention to the fact that, in contrast with what has been shown for the previous VCDU-based delay lines, the outputs of the VCDUs in this oscillator are sign-inverted.

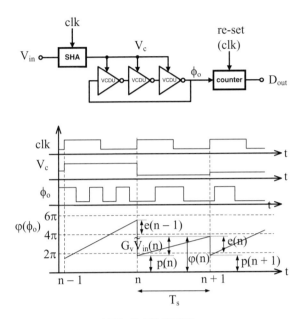

Figure 3.44 A VCO-based TDC [228, 229]. © 2008 IEEE.

counter (which serves the purpose of the quantizer), which is re-set every T_s seconds by the sampling clock, which counts the number of rising or falling edges in ϕ_o during the sampling period T_s, thereby quantizing the VCO phase modulo 2π [229].

The resolution of this quantization is a function of the ratio of the clock period $T_s = 1/f_s$ to the time spacing of the oscillator edges (a function of the number of stages M in the ring oscillator and the VDCU delay τ). So f_s and quantizer resolution are tied: there is a trade-off between the sample rate and the number of quantizer levels [231]. This trade-off, however, improves with CMOS scaling thanks to the decreasing minimum gate delay τ with lower feature sizes.

As shown in Fig. 3.44, the residual phase of ϕ_o at the end of the nth sampling period T_s is the initial phase of the next one:

$$\varphi(n) = p_i(n) + \int_{nT_s}^{(n+1)T_s} K_v \tilde{V}_{in}(n) dt$$

$$= p_i(n) + K_v T_s \tilde{V}_{in}(n)$$

$$= p_i(n) + G_v \tilde{V}_{in}(n)$$

$$= e(n-1) + G_v \tilde{V}_{in}(n) \tag{3.23}$$

where $p_i(n)$ and $e(n)$ represent the initial phase and the residual phase of ϕ_o for the nth sampling period, respectively, K_v is the VCO's frequency gain, and G_v is the VCO's phase gain. The quantized output from the counter, accounting for the 2π periodicity, can be expressed as (apart from some counter latency)

$$D_{\text{out}}(n) = \frac{1}{2\pi} (\varphi(n) - e(n))$$

$$= \frac{1}{2\pi} \left(G_{\text{v}} \tilde{V}_{\text{in}}(n) + e(n-1) - e(n) \right) \qquad (3.24)$$

Taking the z transform of that gives

$$\mathbf{D}_{\text{out}}(z) = \frac{1}{2\pi} \left(G_{\text{v}} \mathbf{V}_{\text{in}}(z) + z^{-1} \mathbf{E}(z) - \mathbf{E}(z) \right)$$

$$= \frac{1}{2\pi} \left(G_{\text{v}} \mathbf{V}_{\text{in}}(z) + (z^{-1} - 1)\mathbf{E}(z) \right) \qquad (3.25)$$

Hence the noise-transfer function (NTF) of this VCO-based TDC is [228, 229]

$$\text{NTF}(z) = \left(z^{-1} - 1 \right) \qquad (3.26)$$

which is equivalent to a first-order $\Delta\Sigma$ ADC. Interestingly, however, this noise-shaped TDC is an open-loop system and, as such, does not require a multi-bit feedback DAC [232].

In passing, we note the similarity of this architecture to ramp-based ADCs [5] and, to some extent, to the previously discussed incremental ADCs (see Section 3.1.5).

Time-interleaving (TI) lowpass $\Delta\Sigma$ ADCs offer a way to obtain a bandpass $\Delta\Sigma$ ADC response [202] much like that provided by "comb filters." This has been theoretically applied to the above VCO-based TDC architecture in [229] using N interleaved VCO-based channels and demonstrated in silicon in [233].

A block diagram of the $N = 8$-channel TI bandpass $\Delta\Sigma$ TDC is shown in Fig. 3.45. The global NTF of this TDC is

$$\text{NTF}(z) = \left(1 - z^{-8} \right) \qquad (3.27)$$

with zeros at

$$z_k = \cos\left(\frac{2\pi k}{N} \right) + j \sin\left(\frac{2\pi k}{N} \right), \quad k = 0, 1, 2, \ldots, N-1 \qquad (3.28)$$

corresponding to the frequencies

$$\omega_k = \frac{2\pi k}{N}, \quad k = 0, 1, 2, \ldots, N-1 \qquad (3.29)$$

The 65 nm CMOS prototype demonstrated in [233] used a master clock at 1.5 GHz; each channel operated at 500 MHz, resulting in an effective sampling frequency of 4 GHz. Owing to the frequency periodicity of the response, the TDC can digitize inputs at DC, 500 MHz, 1 GHz, 1.5 GHz, and 2 GHz. The measured peak SNR and the SNDR at bandwidth 1 MHz were 63.3 dB and 41.5 dB, respectively [233].

We have not yet made explicit mention of the issues limiting the performance of the VCO-based TDC. Besides the classic issues associated with any TI ADC, which will be discussed in Section 3.4 and include gain, offset, timing, and bandwidth mismatches

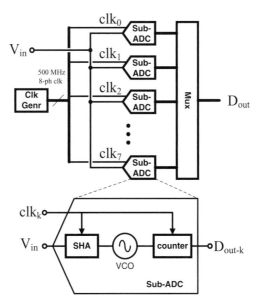

Figure 3.45 A bandpass TDC based on an eight-way time-interleaved architecture made of VCO-based lowpass TDC channels [233]. © 2009 IEEE.

between the interleaved channels that originate all kinds of spurious tonal content, the VCO-based TDC is limited by the following non-idealities, which are discussed in detail in [229] as well as, with proper differences, in an SHA-less version of this architecture reported in [234].

- *Sampling clock jitter*: here the sampling clock is not only used to sample V_{in} with the SHAs (aperture jitter), but also serves as a time reference for integration by re-setting the counters and hence the combined effect is greater than in a conventional ADC; similarly to what happens to conventional ADCs, the phase noise of this clock shows up in the form of skirts around the input signal.
- *VCO phase noise*: the VCO too has its own phase noise, which gets aliased and frequency-shaped by the high-pass NTF of Eq. (3.26).
- *Nonlinearity of the VCO tuning characteristic*: as expected this introduces harmonic distortion into the output ϕ_o of the VCO, partly since this is an open-loop block but primarily because of the fundamental issue that, when one is using a ring-oscillator VCO made of a cascade of VCDU cells, it is the oscillation period T that is proportional to the control voltage V_c, not the oscillation frequency [235]. So there is a strong nonlinearity (reciprocation) in this characteristic without even accounting for deviations from linearity in the $\Delta T(V_c)$ characteristic of Fig. 3.37. The ring-oscillator VCO nonlinearity introduces a strong second-, third-, and fourth-order distortion, with weaker higher-order spurs.[15] In this example, to mitigate this effect, a frequency

---

[15] It has been reported that in a typical VCO with 5% tuning nonlinearity an ENOB of the order of 6 is to be expected without compensation [231].

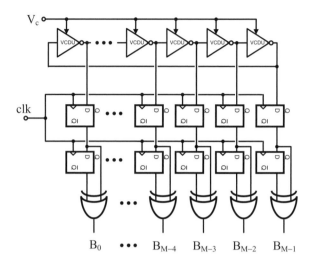

Figure 3.46 An alternate VCO-based quantizer [231]. © 2008 IEEE.

planning based on the use of several channels and proper choice of the passband frequencies has been applied in order to have a lot of the corresponding spurious content
fall out of the passband.

- *Considerations on the SHA*: unlike conventional ADCs, in this TDC the input is
  sampled but not directly quantized, while it is instead integrated by the VCO before
  quantization. It can be seen that non-zero settling time or a steady-state error of
  the SHA does not matter, but its linearity is important. To remove memory effects,
  a return-to-zero (RTZ) SHA has been used in [229]. Other issues include charge
  injection, clock feedthrough, and slew-rating.

A variation of the VCO-based quantizer is depicted in Fig. 3.46 [231]. Here the
outputs of all the M VCDUs are directly fed to a set of M DFFs (an M-bits register),
which quantizes the VCO outputs at the sampling time and stores their state. Another
identical M-bits register is cascaded to the latter (storing the previous sample), and the
outputs of the two registers are combined by means of M XOR gates. It can be shown
that this alternate VCO-based quantizer of Fig. 3.46 has the same first-order high-pass
noise-transfer function $\text{NTF}(z) = 1 - z^{-1}$ as that of the architecture of Fig. 3.44 [231].
One important advantage of this alternate quantizer is that the outputs of the M XOR
gates correspond essentially to a thermometric representation of the quantizer output,
and this thermometric code circulates as in a barrel shifter; namely the position of the
next sets of ones will begin where the previous set of ones ended, as shown in Fig. 3.47
[231]. That is a powerful feature applied in the architecture that will be described next.

As mentioned above, an important limitation of the VCO-based quantizers is the
harmonic distortion introduced by the VCO tuning characteristic. This distortion can
be reduced by means of negative feedback if, for example, the quantizer is used in a
$\Delta\Sigma$ modulator, as shown in Fig. 3.48 [228, 231]. In particular, a CT $\Delta\Sigma$ modulator is
discussed in [231]. If the barrel-shifter thermometric M-bit output of the quantizer is

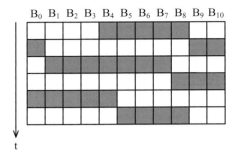

Figure 3.47 The outputs of an 11-unit VCO-based quantizer over six consecutive samples and their barrel-shifter feature [231]. © 2008 IEEE.

Figure 3.48 A continuous-time $\Delta\Sigma$ using the VCO-based quantizer.

directly used to drive the M individual, nominally identical, current-steering cells of the fully thermometric feedback DAC, then dynamic element matching (DEM)[16] has implicitly been introduced to scramble the mismatches in the M units of the DAC without requiring additional hardware, power, or delay. Furthermore, the VCO-based quantizer effectively increases the order of the noise shaping by 1, in addition to what is provided by the loop filter $H(s)$ thanks to its intrinsic NTF$(z) = 1 - z^{-1}$. The linearity of the VCO tuning characteristic is enhanced by the feedback action and, in particular, by the gain of the loop filter $H(s)$ placed in front of it. A 0.13 μm CMOS prototype discussed in [231] demonstrated a measured performance of SNR $= 86$ dB and SNDR $= 72$ dB with a 10 MHz signal bandwidth, clocked at 950 MHz, consuming 40 mW at a 1.2 V supply, using a ring-oscillator with $M = 31$ VCDU cells and a second-order $H(s)$ (hence achieving overall a third-order noise shaping thanks to the additional zero in the VCO's NTF).

Let us note that for both of the VCO-based quantizers of Figs. 3.44 and 3.46, on quantizing the oscillation edges within the sampling period T_s, frequency is what is actually quantized. Again, there is a nonlinear relation between the *frequency* of oscillation of the VCO and its control voltage V_c, while the relation between the VCO's *phase* and the control voltage V_c is considerably more linear [236]. A further improvement on the design reported in [231] deriving from this observation has been described in [236]. The CT $\Delta\Sigma$ TDC 0.13 μm CMOS prototype with VCO-based quantizer reported in [236] demonstrated a measured performance of SNR $= 81.2$ dB and SNDR $= 78.1$ dB with a 20 MHz signal bandwidth, clocked at 900 MHz, consuming 87 mW at a 1.5 V supply.

---

[16] In fact, specifically, this is identical to data-weighted averaging (DWA) [195].

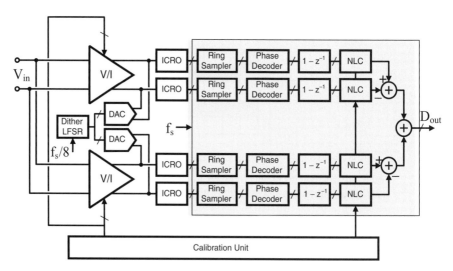

Figure 3.49 A digitally linearized VCO-based quantizer [235]. © 2010 IEEE.

Another approach by which to tackle the linearity issue of VCO-based quantizers has been reported in [235]. This uses a combination of techniques, including digital background correction of the VCO nonlinearity and self-canceling dither. In this approach the VCO-based quantizer is again treated as an open-loop TDC. A simplified block diagram of the architecture described in [235] is depicted in Fig. 3.49. Four VCO quantizer paths are visible. Each of them is composed by a fully differential 15-unit ring-oscillator made of *current-controlled* delay units (ICDUs), followed by a phase decoder, an (implicit) differentiator $1 - z^{-1}$, and a nonlinearity correction (NLC) block. In the following we will refer to the current-controlled ring-oscillator as the ICRO.

Although the ICRO is fully differential, as previously discussed, its nonlinearity contains significant even-order distortion terms. Therefore, by using a second, nominally identical, path so that the two paths combined work in pseudo-differential fashion, the difference output has a dramatically lower (limited by path matching) even-order distortion and a net 3 dB improvement in SNR (the difference signals have twice the amplitude and hence 6 dB more power than in a single path, and the combined uncorrelated noise adds quadratically and hence has 3 dB more power than in a single path; overall the SNR improved by 6 dB signal minus 3 dB noise, hence there is a 3 dB net improvement).

Dither is differentially injected into the control inputs of the two ring-oscillators using a current-steering DAC and helps to improve the linearity of the combined two-path structure at the expense of some signal range. To cancel out the dither from the output signal, a second pair of pseudo-differential ICRO quantizer paths is needed. This second pseudo-differential path is fed with the same differential input and an opposite differential dither so that when the two difference outputs are added together the dither is canceled out (to the extent of the channels' matching), while the net differential signal amplitude doubles again (once again, another 3 dB SNR improvement is expected in principle).

Note that, compared with a single non-differential path, the four signal paths consume four times the power (6 dB increase) and circuit area. However, they result in a net $3\,dB + 3\,dB = 6\,dB$ SNR improvement. Considering the thermal FOM_2 of Eq. (2.11) or the FOM_3 of Eq. (2.12), a 1 dB increase in power consumption is expected for a 1 dB increase in dynamic range. Hence, to the extent that the SNDR is noise-limited, the use of multiple paths does not impact the $FOM_{2/3}$ [235].

The matching of the pairs of paths constituting the pseudo-differential path is not sufficient to cancel out the harmonic distortion to better than an SFDR of the order of 60–65 dBc [235]. That is one of the reasons behind the NLC block: the nonlinear tuning characteristic of the ICRO has been modeled and has been corrected by the look-up-table-based NLC block. The coefficients for this correction are determined in the background using yet another replica of one of the quantizer paths. This replica is fed with three two-level, independent, zero-mean, pseudo-random sequences used to determine the correction coefficients, and the latter are then applied to the NLC. The coefficients are periodically updated every $2^{28}/f_s$ seconds. The replica is also used to calibrate the center frequency of the tuning range of the ICROs.

A 65 nm CMOS prototype was reported in [235]. It had a measured performance of $SNR = 70\,dB$, $SNDR = 67\,dB$ with an 18 MHz signal bandwidth, clocked at 1152 MHz, consuming 17 mW. An attractive aspect of this architecture is that it is all digital, except for the voltage-to-current converters used to drive the VCOs.[17]

3.2.4   Digitally assisted analog techniques

As discussed in Section 1.2, the relentless march toward finer-scale CMOS technologies is proving to be both challenging and full of opportunities for analog and mixed-signal designers [239]. Not only do the cost, availability, and speed of digital gates improve on going from one CMOS node to the next finer one, but also the energy efficiency for a given logic function has been significantly improving. For example, it has been noted in [82] that a two-input NAND gate dissipates roughly 1.3 pJ per logic operation in a 0.5 µm CMOS process while the same gate dissipates only 4.5 fJ in a more recent 90 nm process. That is approximately a 300-fold improvement in only 10 years. Although ADCs too have become increasingly efficient (as will be discussed in Section 5.1), these and other analog and mixed-signal devices have certainly not seen such a dramatic energy improvement: a mere ∼32-fold improvement in energy efficiency has been assessed over the same decade [82].

Therefore, in order for converters to evolve and thrive (almost in a Darwinian sense of the term) in this changing environment, the question of whether these devices should take full advantage of what digital technology can offer forcefully leads even the most conservative and recalcitrant analog designers to an inevitable positive answer [239].

Indeed, there are various degrees and different ways in which one may want to make use of this evolution, and a few such approaches have been explored in this chapter.

---

[17] And for the DACs used to inject the dither, although it should be said that these DACs do not need to be high-performance converters.

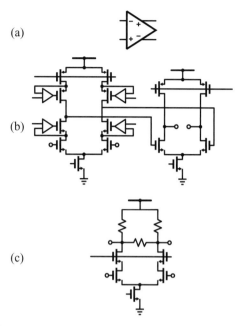

Figure 3.50 Differential amplifiers: (a) a generic fully differential amplifier; (b) a traditional two-stage amplifier capable of a DC gain greater than 100 dB in a standard 0.35 μm CMOS process; and (c) a resistively loaded differential pair designed for a DC gain of 8 [240].

The techniques briefly discussed in this section originate from the broad philosophy of designing analog circuits knowingly lacking the required precision for the target performance specification because the resulting inadequacies are compensated for using digital processing.

Let us consider an example to fix the ideas. Let us consider the multi-bit front MDAC stage of a 14 b pipelined ADC. That is specified to have an op-amp with an open-loop DC gain in excess of 100 dB so that the closed-loop gain accuracy of this switched capacitor stage is met for 14 b [241] (see Fig. 3.50(a)). Achieving such a high open-loop DC gain with a very low supply (think less than 1.5 V) and low transistor intrinsic gain g_m/g_{ds} (think less than 15) while, at the same time, using no more than two cascaded stages for wide closed-loop bandwidth and stability (such as that shown in Fig. 3.50(b)) is certainly non-trivial. But these are realistic conditions in 90 nm and beyond.

So, the alternate approach proposed in [240] is to modify this MDAC stage to use a very simple open-loop amplifier (a resistively loaded differential pair such as the one shown in Fig. 3.50(c), in this specific example) designed to realize the closed-loop gain of the residue MDAC stage (8 in this specific example). Such an amplifier is much more amenable to a lower supply and, due to its simpler topology and open-loop configuration, is also potentially much faster than the previous one. On the other hand, since the new amplifier is used in open-loop configuration, a lot of the crucial benefits of feedback circuits are lost and need to be dealt with. Specifically, the new circuit is much more sensitive to PVT variations, both in terms of the nominal values of its key design parameters (gain, bandwidth, biasing, etc.) and in terms of how these vary during the normal operation of the circuit. Moreover, it is considerably nonlinear.

That is where digital resources come to the rescue. The idea is that by carefully modeling the non-idealities introduced by this more CMOS scalable architecture one can correct for the analog inadequacies by properly digitally processing the converted output of the ADC. In other words, by digitally calibrating for the gain, offsets, and, more importantly, harmonic distortion introduced by the new op-amp, it is possible to digitally post-process the converted data and recover from the analog nonlinearities.

In this specific example, the amplifier of Fig. 3.50(c) was carefully designed to have a predominantly cubic distortion while having negligible second and fifth harmonics. That was carefully modeled, and a cancelation algorithm that relies on signal statistics (a least-mean-square-, LMS-, based calibration) to measure the distortion coefficients then digitally corrects for the nonlinearity. It should also be said that in order to obtain sufficient accuracy for the measurement of errors and subsequent digital calibration two more bits of resolution are required on the output of the ADC. Two additional (LSB) bits are then added to the nominal output bits and used to achieve this required higher internal resolution. The power penalty for this addition is, however, fairly minor since these redundant comparators make only a small contribution to the overall ADC power.

Some possible misunderstandings should be avoided from the outset. The criticality of the work of the analog designer is not diminished by the fact that some coarse circuitry has replaced a more traditional "precision" approach. On the contrary, it is crucial that the analog designer does the following.

- The designer must carefully choose the proper "scaling-friendly" structure (e.g. a resistively loaded differential pair in the above example).
- The designer must model carefully the overall circuit with the simpler active circuitry embedded in it; for example, what is the proper mathematical representation of the new stage? What needs to be accounted for and how does it play? Is it primarily a static cancelation that is required or do dynamic effects play a significant role?
- The designer has to minimize the number and the type of non-idealities that result from this choice. For example, the same architecture as above would lead to a strong second- and fifth-order distortion for a lower overdrive voltage of the differential pair transistors, making the digital cancelation of these additional terms a new challenge and leading to higher complexity.
- The designer must find a way to effectively detect the errors and correct for them. For example, is a test signal required for extracting the errors from the digitized output or can the input signal itself be used to detect them? Will the detection converge to the identified error coefficient quickly enough or does it require observing over very long time series while the parameters to be corrected have already changed enough without having been compensated for? Will the detection and correction hardware

require more power and area than what was originally aimed to be saved by doing away with the precision analog circuitry? Can the errors actually be detected in the first place or do they become indistinguishable from others, or do they experience further irreversible distortion before they can be detected?[18]

In other words, while "scaling-unfriendly" precision circuits might be replaceable with other type of circuitry, the need for deeply knowledgeable analog designers is even more critical when this path is taken. To get a sense of the potential efficiency improvement, a direct comparison has been reported in [240], where, for the above-cited example consisting of replacing the first stage of a 14 b 75 MSPS pipelined ADC with a modified one using an open-loop $G_m R$ amplifier, the corresponding power consumption was reduced by approximately 60%–70%. Indeed, the effectiveness of this approach is expected to become particularly more compelling in deeper scaled CMOS technologies.

This example, as previously mentioned, was presented primarily to better explain the key concepts. Namely, accepting that analog, mixed-signal, and RF circuits face a number of shortcomings when designed for fine-line CMOS processes, the availability and efficiency of digital processing can be exploited to correct errors and "assist" (hence the name) in addressing them [242].

Such a concept has actually been more or less "conscious" in its essence for a long time. For example $\Delta\Sigma$ ADCs [81, 82, 175, 243] could be considered a precursor of this idea in the sense that, for instance, large oversampling is a way to exploit the intrinsic high speed of a CMOS technology to obtain higher accuracy (mainly lower noise) in a narrow analog bandwidth. Also, dynamic element matching (DEM) is yet again a way to leverage digital processing and oversampling to improve the linearity of the DACs even without measuring it. Despite the further addition of digital decimation filters, overall, such architectures are among the most accurate and power-efficient when speed requirements can be met [81].

Also, the general concept of intentionally designing smaller and faster analog circuits and then digitally calibrating/correcting the resulting linearity errors originating from component mismatch or the parameter sensitivity originating from PVT variations has been pointed out in numerous contexts as an enabling capability of calibration [73, 98, 99, 100, 101, 129, 244, 245, 246, 247, 248].

An interesting recent evolution of this concept is associated with adaptive and reconfigurable ADCs [249]. The latter are converters that can be digitally controlled to vary the performance or the interconnection of some of their building blocks, therefore changing their overall conversion specifications in response to different "modes of operation." For example, in a unique wireless system that is meant to be used for different communication standards (e.g. WLAN, DVB, UMTS, Bluetooth, ...) an ADC can be "programmed" to have, say, one mode of operation with a higher dynamic range and narrower signal band and another mode of operation with a lower dynamic range and broader band [250]. In

---

[18] In mathematical terms, if a nonlinear distortion $y = f(x)$ is what one wants to reverse for, does $x = f^{-1}(y)$ exist or, at least, does it exist for the narrow operating range of x and y?

doing so, the operating conditions and power consumption of the ADC are optimized for the intended mode of operation. The idea can, in fact, be further developed if, for example, the spectrum of the input signal is being sensed and the converter is automatically reconfigured for optimal performance [251]. Another interesting approach, still in the context of communication systems, was proposed in [252], where the pilot tones of an OFDM system (normally used for communication-channel equalization) were exploited to extract mismatch information on the ADC (taking advantage also of the existing FFT block in the back-end communication processor), and the latter was then calibrated and operated optimally in the communication system. That is an example of a digitally assisted analog system. But it is also an example of a more synergistic approach whereby the ADC is not looked upon as a universe on its own, but rather as part of a more complex system where DSP resources can be available to digitally enhance various other parts of the analog sub-system.

3.3   Hybrid ADCs

Various ADC architectures have been discussed throughout this chapter and their similarities have been highlighted where appropriate. But hybrid ADCs are those architectures that result from the combination of multiple others: a converter that is a mix of a $\Delta\Sigma$ and a pipelined ADC [253], or another that is a mix of an SAR and a pipelined ADC [166], or one that implements the binary search principle of an SAR ADC as part of a two-step converter [165], and so on.

Hybrid architectures are often the result of the designer's intent to combine the strengths of different types of ADCs and actually create an architecture that, at least for a given set of specifications, is overall better than any one of its components.

For instance, $\Delta\Sigma$ ADCs tend to be more suitable for higher resolution and lower signal band than pipelined ADCs (which, in turn, are well suited to wideband inputs but are usually a better fit to moderate resolution). Combining a $\Delta\Sigma$ front end with a pipelined ADC back end was the triggering idea for the 16 b cascaded (MASH) hybrid ADC discussed in [253] to obtain high resolution (89 dB SNR) *and* high speed (1.25 MHz input bandwidth) (respectably so even after 15 years) without requiring calibration.

The block diagram of this cascaded (MASH) converter is shown in Fig. 3.51. Here a first k-bit $\Delta\Sigma$ modulator is cascaded with a high-resolution m-bit second stage. This is similar to a Leslie–Singh [81] architecture where only some of the quantizer bits are fed back to the input of the modulator. The second-stage ADC provides m-bits at the same rate as the $\Delta\Sigma$ modulator's quantizer. In this structure, since the m-bit quantizer is not closed in a loop, any associated latency is unimportant. The D-cycle delay z^{-D} is used to equalize the $\Delta\Sigma$ modulator output with the m-bit quantizer output which is, in fact, implemented using a pipelined ADC (hence having latency).

A more detailed block diagram of this 16 b hybrid $\Delta\Sigma$-pipelined ADC is shown in Fig. 3.52. In this diagram the $\Delta\Sigma$ modulator and the pipelined ADC are much more explicitly visible on the left- and right-hand side of the diagram, respectively. The first stage of the pipelined ADC is, in fact, directly used by the $\Delta\Sigma$ modulator as a

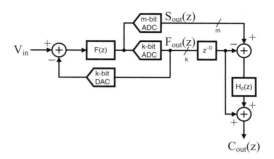

Figure 3.51 A cascaded (MASH) $\Delta\Sigma$ ADC composed of a first $\Delta\Sigma$ k-bit stage and a second m-bit quantizer stage. The output bits are combined and decimated, leading to the digital output [253]. © 1997 IEEE.

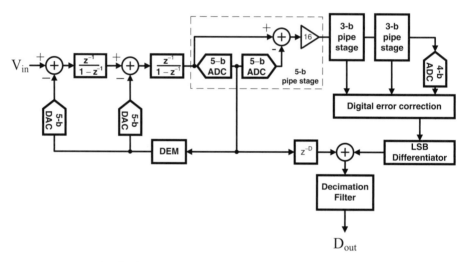

Figure 3.52 A more detailed block diagram of the hybrid 16 b $\Delta\Sigma$-pipelined ADC [253]. The second-order multi-bit $\Delta\Sigma$ modulator is clearly visible on the left-hand side and its 5 b quantizer is shared with the sub-ADC of the first residue stage of the pipelined ADC. The pipelined ADC can be seen on the right-hand side of this block diagram and is composed of a first 5 b stage, two subsequent 3 b stages, and a final 4 b flash ADC. Each stage has one redundant bit for digital error correction [253]. © 1997 IEEE.

quantizer as well as being part of the first residue stage of the pipelined ADC. The $\Delta\Sigma$ modulator is a traditional second-order loop with a 5 b quantizer. The linearity of the two feedback DACs is improved using DEM. In order to have a wideband input, the oversampling ratio of the modulator is rather low[19] (8×). Hence the input band is BW $= f_s/(2 \cdot \mathrm{OSR}) = 20\,\mathrm{MHz}/(2 \cdot 8) = 1.25\,\mathrm{MHz}$. The pipelined ADC is a

---

[19] The sampling rate is limited by the speed of the process, hence the need to resort to a multi-bit loop in the modulator and, overall, to a cascaded architecture.

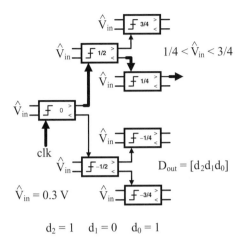

$\hat{V}_{\text{in}} = 0.3 \text{ V}$

$$d_2 = 1 \quad d_1 = 0 \quad d_0 = 1$$

Figure 3.53 A 3 b CABS ADC. There is a layer for each of the n bits to resolve and only one comparator per layer is triggered (turned on), while all others are "dormant." The sampled input signal $\hat{V}_{\text{in}}$ drives the binary search algorithm in an asynchronous domino-style cascade [165]. © 2008 IEEE.

four-stage 12 b 20 MHz ADC, where each multi-bit stage has one redundant bit, for digital correction. Second-order digital differentiation of the LSBs, which is provided in the LSB differentiator block, emulates the filtering of the flash quantization noise in the second-order modulator loop [253].

The careful selection of the architecture and its parameters in this hybrid provides its performance: a 12 b pipelined ADC does not require calibration, the combination of a MASH structure and a multi-bit loop in the front-end $\Delta\Sigma$ allows the use of a limited OSR. Overall the converter has 16 b resolution and achieves 89 dB SNR in a bandwidth of 1.25 MHz. It consumed 550 mW and was implemented on a 0.6 µm CMOS technology with dual $5AV_{dd}$ and $3DV_{dd}$ supplies.[20]

A more subtle form of hybrid ADC was reported in [165]. This architecture, called "comparator-based asynchronous binary search" (CABS), can be thought of as a hybrid of a flash ADC and an SAR ADC. The CABS architecture is depicted in Fig. 3.53.

All the comparators in the CABS of Fig. 3.53 have a built-in reference voltage against which they compare its input. That is, for example, the calibrated dynamic comparator discussed in Section 3.1.2 and shown in Fig. 3.7.

In this ADC the sampled input $\hat{V}_{\text{in}}$ is simultaneously applied to all the $2^n - 1$ comparators, just like in a flash ADC. However, unlike in a flash ADC, where all $2^n - 1$ comparators are simultaneously operating and digitizing the input, in this architecture only n comparators are turned on. Therefore a CABS ADC has a greater energy efficiency than that of a flash ADC.

---

[20] These last few numbers should be put into context by comparison with the standards of performance of 1997!

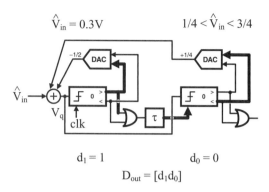

$\hat{V}_{in} = 0.3\,\mathrm{V}$ $1/4 < \hat{V}_{in} < 3/4$

$d_1 = 1$ $d_0 = 0$

$D_{out} = [d_1 d_0]$

Figure 3.54 A 2 b hybrid SAR ADC [165]. © 2008 IEEE.

The principle of operation of the CABS architecture, in turn, is essentially the same as that of an SAR ADC, namely the sampled input $\hat{V}_{in}$ drives a chain of decisions based on a binary search algorithm that ultimately finds the range (the quantum) within which $\hat{V}_{in}$ lies. As the search progresses, just like in an SAR ADC, the binary representation of $\hat{V}_{in}$ is determined bit by bit, beginning from the MSB and progressing toward the LSB.

An example will clarify the operation of CABS. Let us assume $\hat{V}_{in} = 0.3$ V. Initially all comparators are turned off and all their outputs are zero. At the rising edge of the clock the first comparator on the left-hand side is activated. This compares the input $\hat{V}_{in}$ with its built-in reference of 0 V and determines the MSB d_2. If the input is larger than the reference ($\hat{V}_{in} > 0$ V) then $d_2 = 1$ and the upper comparator of the next layer is activated. Conversely $d_2 = 0$ and the lower comparator of the next layer is triggered. In this example, the former case is true and the upper comparator of the next layer compares $\hat{V}_{in}$ with its own built-in reference of $1/2$ V. The behavior of this step is the same as that of the previous one. Namely, since $\hat{V}_{in} = 0.3$ V $< 1/2$ V, $d_1 = 0$ and the lower comparator of the next layer is triggered. Finally, since $\hat{V}_{in} = 0.3$ V $> 1/4$ V, $d_0 = 1$ and the conversion is over. The falling edge of the clock turns all comparators off again to a "dormant" state, until the next conversion.

Note that only $n = 3$ comparators were activated. Each comparator triggered the next one asynchronously like tiles in a domino chain. CABS has also been referred to as a "binary search" ADC elsewhere in this book. It is also important to remark that, unlike with a traditional SAR ADC, there is no need for a controller directing the binary search or for an internal high-frequency clock generator timing the operation. All of the circuitry used is purely dynamic (see Section 3.1.2 on dynamic ADCs) and so this converter is extremely power-efficient. However, as in flash ADCs, the physical size of the converter grows exponentially with the number of bits.

Another hybrid was proposed in the same paper [165]. That too is similar in principle to an SAR ADC or to a two-step converter. A 2 b ADC case is depicted in Fig. 3.54 to illustrate its operation. This time all of the comparators have a built-in reference equal to zero. The sampled input is fed to the summing node and initially both comparators are off and with their outputs at zero. Therefore the two 1-bit DACs have zero output as well and $V_q = \hat{V}_{in}$. Let us again assume $\hat{V}_{in} = 0.3$ V. The rising edge of the clock triggers the first comparator, which senses $V_q = \hat{V}_{in} = 0.3$ V. That is greater than its built-in reference ($V_q > 0$ V) and hence it sets $d_1 = 1$ and turns on the first DAC. The DAC subtracts $1/2$ V from the input at the summing node and V_q changes to $V_q = 0.3 - 1/2 = -0.2$ V. The first comparator is now locked, so it does not change state any more (until it is re-set), but its output also goes through the OR gate and the delay τ. So after τ seconds the second comparator is triggered. τ is chosen to be long enough for the first DAC and V_q to settle. So the second comparator now sees $V_q = -0.2$ V (the residue of the prior operation) and compares it with its built-in reference (zero). Since $V_q = -0.2$ V < 0 V, the output of this second comparator is opposite to that of the previous one and it sets $d_1 = 0$. The second DAC is not activated and its output stays at zero. A similar OR gate and delay τ could trigger another such stage and continue resolving more bits.

This architecture operates an asynchronous binary search similarly to an SAR ADC (or a pipelined ADC, for what matters), has only n comparators, and does not require a controller to direct its operation. However, unlike the CABS architecture, it does not grow exponentially in complexity with the number of bits n.

A hybrid of these two hybrids was then presented in the same paper [165] by replacing the second comparator of the architecture of Fig. 3.54 with a 6 b version of the CABS ADC of Fig. 3.53. This interesting two-step hybrid has then a 1 b coarse first step and a 6 b fine second step and resolves 7 b with $1 + (2^n - 1) = 1 + (2^6 - 1) = 2^6 = 64$ dynamic comparators (as opposed to the 127 comparators that would be required if a 7 b flash ADC were used). A 7-bit implementation of this hybrid ADC in a 90 nm digital CMOS process with a 1 V supply was reported to achieve 40 dB SNDR at 150 MS/s, consuming 133 μW [165].

The same authors also proposed other similar hybrids in [254] and [255]. In [254] a new type of folding stage is placed in front of a flash ADC built using the dynamic comparators with built-in reference of Fig. 3.7. The folding stage resolves the MSB, effectively halves the input range by folding it onto its middle, and then resolves the remaining bits using the flash ADC.

In [255] the same folding stage is placed in front of a 3 b binary search ADC just like that of Fig. 3.53, and the latter is followed by a 2 b flash ADC that uses the dynamic comparators with built-in reference of Fig. 3.7. The resulting hybrid ADC has a resolution of 1(folding) + 3(CABS) + 2(flash) = 6 b. Then four of these hybrid converters are time-interleaved. A 6 b 2.2 GSPS prototype of this ADC was implemented in a 40 nm digital CMOS with a 1.1 V supply and had a low-frequency SNDR of 31.6 dB, consuming 2.6 mW [255].

The last example of a hybrid ADC covered in this section is a two-stage pipelined ADC where each stage is a multi-bit SAR ADC [166]. The aim of this high-resolution (18 b) high-speed (12.5 MSPS) hybrid is to conjugate the high accuracy and energy efficiency of SAR ADCs with the higher conversion speed and the ability to relax the sub-ADC's accuracy of pipelined ADCs. Furthermore, in order to speed up the conversion rate of the sub-SAR ADCs, multi-bit SARs resolving 2 b per conversion iteration are used.

The block diagram of this hybrid ADC is shown in Fig. 3.55. Although the architecture of the two sub-SAR ADCs is slightly different, as will be shown in the

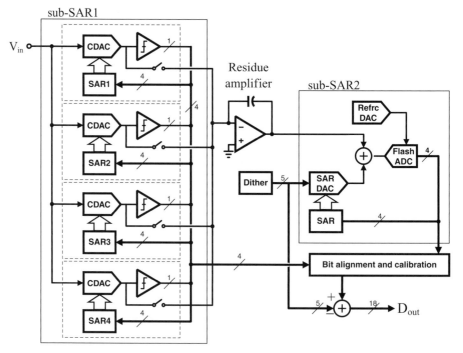

Figure 3.55 An 18 b 12.5 MSPS hybrid pipeline/SAR ADC [166]. © 2010 IEEE.

following, sub-SAR1 samples the input V_{in} and resolves 10 b by performing five successive approximation iterations, each one of which resolves 2 b per cycle. These are the nine most significant bits, plus one redundant bit to be used for error correction. The converter sub-SAR1 also couples with a residue amplifier (RA) synthesizing the conversion residue $\hat{V}_{res}$ ($\hat{V}_{res} = \hat{V}_{in} - \tilde{V}_{in}$, where $\tilde{V}_{in}$ is the quantized analog representation of $\hat{V}_{in}$), and this residue is fed to the second converter, sub-SAR2, which quantizes it, completing the conversion by resolving nine more bits (with one bit of overlap across the RA output) over two successive approximation cycles, each one returning 5 bits.

To correct for capacitor mismatch in the DACs, the outputs of sub-SAR1 and sub-SAR2 are digitally corrected during the conversion (see the block "bit alignment and calibration" in the lower right corner of Fig. 3.55). The error-correction coefficients are determined at final factory test and stored in a non-volatile memory. The same "bit alignment and calibration" block also performs the combination of coarse and fine bits to provide the final conversion result.

In more detail, sub-SAR1 is composed of four SAR ADCs, each one of which has a capacitive DAC (CDAC). Each CDAC samples V_{in} and then presents its comparator with $\hat{V}_{in} - V_{refk}$, where V_{refk} ($k = 1, 2, 3, 4$) is the reference voltage corresponding to the successive approximation algorithm. The references V_{refk} are evenly spaced between each SAR so that the four comparators sense four levels and, combined together, actually behave as a four-level flash ADC: three levels suffice to resolve 2 b; the fourth level is redundant and allows digital error correction.[21] All the results from the four comparators are fed to the four SAR controllers and used to set the CDACs for the next approximation iteration. The converter sub-SAR1 then works as a multi-bit SAR resolving 2 b per cycle (with redundancy) in five approximation iterations and hence, in the end, returns 10 b that are the nine MSBs plus one redundant bit.

At the end of the fifth iteration, the CDACs are reconfigured to all have the same output, which corresponds to the residue of the conversion and, by charge transfer (the four switches shown next to the four SAR comparators close) into the feedback capacitor of the residue amplifier (RA), depicted in the center of Fig. 3.55, the residue is amplified[22] by 42.66 and fed to sub-SAR2 to be further quantized. Also, when this coarse conversion phase has been completed, sub-SAR1 is ready to sample and quantize the next input; that happens while sub-SAR2 quantizes the residue.

Converter sub-SAR2, shown on the right-hand side of Fig. 3.55, does not need to perform a residue amplification, hence it can have a much simpler architecture. It is still a multi-bit SAR that compares its input against a four-level ADC (a flash ADC with autozeroed comparators). Again, that is 2 b plus one more redundant level for error correction. In sub-SAR2 the SHA function is implemented by the RA while the SAR DAC merely synthesizes the quantized approximation of the input. The flash ADC reference is provided by a separate reference DAC, which also scales the flash ADC's range during the conversion process to account for the resolution of multiple bits per approximation cycle.

Another function of the SAR DAC is to inject dither into the conversion. The dither is used to improve the differential linearity (DNL) of sub-SAR2 and it is digitally subtracted from the output as shown on the summation block depicted on the lower right-hand side of Fig. 3.55. The result of the fine conversion from sub-SAR2 and that of the coarse conversion from sub-SAR1 are combined by the digital encoder on the bottom left of Fig. 3.55, which also takes care of performing the digital error correction. As sub-SAR2 completes its digitizing function, sub-SAR1 is ready to provide the next residue and coarse-conversion result.

An 18 b 12.5 MSPS prototype implemented in a 0.25 μm CMOS process achieving 93 dB SNR for a 50 kHz input tone and consuming 105 mW is reported in [166]. A converter with these specifications finds application in digital X-ray systems, where it is actually switched between different digitization channels. Similar dynamic performance could be achieved by traditional SAR ADCs, but these have a much longer conversion time due to the many approximation iterations required. Similarly, traditional $\Delta\Sigma$ ADCs can also have excellent dynamic performance, but the transient response associated

---

[21] The fourth redundant level is actually not used during the first SAR cycle; however, that's a detail.

[22] The choice of this gain accounts for the redundancy/range scaling as well as the practicality of capacitor ratios. Also, the RA is autozeroed to reduce offset and $1/f$ noise while its gain is factory trimmed. Just like in traditional pipelined ADCs on gaining up the residue, the offsets of the comparators in the following stage now relate to a magnified residue; therefore, by comparison, they are equivalent to much smaller offsets when reported to the RA's input. Specifically, these offsets have been equivalently gained down by the amount of the residue amplification: a factor of 42.66.

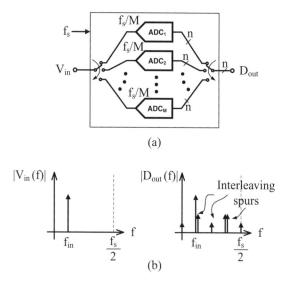

(a)

(b)

Figure 3.56 (a) A generic n-bit M-channel TI ADC with sample rate f_s (each sub-ADC samples at f_s/M) [258]. © 2001 IEEE. (b) The spectrum of a sine-wave input (on the right-hand side) and the resulting spectrum of the TI ADC output in the presence of channel mismatches (in this example, $M = 4$). Adapted from [261].

with the modulator and the decimation filters is too long to allow acceptable channel switching rates [166].

3.4   Time-interleaving

Time-interleaved ADCs (TI ADCs) have been known for decades [256], and have allowed digitization of signals with extremely high sample rates, medium to low dynamic performance, and reasonable power consumption given the resulting speed. For example, in [116] a large (80 ADC channels!) single-chip TI ADC array implements an 8 b 20 GSPS ADC in a 0.18 μm CMOS process consuming overall about 9 W (ADC buffer not included) and performing at 6.5 ENOB with 500 MHz input and 4.6 ENOB with 6 GHz input. Also, in [162], eight channels are interleaved in a single-chip TI ADC implementing a 6 b 600 MSPS ADC in a 90 nm CMOS process consuming 10 mW and performing at 5.2 ENOB with 100 MHz input and 3.8 ENOB with 329 MHz input. In a more recent example, a 48-channel 10 b 2.6 GSPS TI ADC on a 65 nm CMOS process consumes 480 mW for a 7.76 ENOB performance up to the Nyquist frequency [257].

A generic block diagram of an M-channel TI ADC is shown in Fig. 3.56. Each of the M sub-ADCs (channels) samples and converts at $1/M$ of the total sample rate f_s, digitizing n bits. The sampling occurs in repetitive sequence every $T_s = 1/f_s$ seconds: ADC1 samples $V_{in}(t_0)$ first, then, T_s seconds later, ADC2 samples $V_{in}(t_0 + T_s)$, then

ADC3 samples $V_{\text{in}}(t_0 + 2T_s)$, etc. Finally, ADC$M$ samples $V_{\text{in}}(t_0 + (M-1)T_s)$ and the cycle repeats when ADC1 samples $V_{\text{in}}(t_0 + M \cdot T_s)$ and so on. The M n-bit outputs are time de-multiplexed so that an n-bit sequence D_{out} at sample rate f_s is recovered in the correct sampling order.

A very appealing aspect of TI ADCs is that, in principle, the sample rate f_s increases linearly with the number of channels M, as do the power consumption and area. Conversely, using a single ADC, for example a pipelined ADC, generally the sample rate is increased by increasing the transconductance G_m of the amplifiers so that they can settle faster to the required accuracy. Unfortunately, increasing the bias current to increase the transconductance reaches a point of diminishing returns due to the nonlinear law relating G_m and the bias current. Moreover, as device sizes grow, so do their parasitics. Ultimately, for a given process technology, pumping more current into the amplifiers first becomes expensive in power and area and eventually simply stops bringing any further speed advantage.

Because of this linear relation between f_s and M and the power and area of the TI ADC, researchers have recently found the combination of interleaving and SAR ADCs particularly powerful, the reason being that SAR ADCs require relatively small area and are very power-efficient. Although these are not the fastest architectures, interleaving a large number of them can lead to high sample rates with excellent energy efficiency [159, 162, 163, 257].

Unfortunately, the biggest challenge with interleaving, and the reason why this approach hasn't spread out in a viral way, is that dynamic performance is severely limited by channel mismatches. Mismatches between the sub-ADCs affect the different quantized outputs $\tilde{V}_{\text{in}}(t_0)$, $\tilde{V}_{\text{in}}(t_0 + T_s)$, ..., $\tilde{V}_{\text{in}}(t_0 + (M-1)T_s)$ and so on consistently with the mismatched characteristics of the channels. So, when these digitized outputs are recombined to produce D_{out}, the input digitized series has been convolved with the mismatched channel characteristics in a cyclic fashion. Specifically, the main channel-to-channel mismatches playing a role are the following.

- *The sub-ADC offset error mismatch*: each sub-ADC ADC$_j$ has its own offset offset$_j$, and it adds that to the samples it digitizes. Therefore an M periodic sequence offset$_j$, $j = 1, 2, \ldots$ with period equal to M/f_s is superimposed on the digitized signal and shows up as tones at $f_o = k \cdot f_s/M$, for $k = 0, 1, 2, \ldots$, with power that depends on the amount of offset mismatch only.
- *The sub-ADC gain error mismatch*: each sub-ADC ADC$_j$ introduces its own small gain (frequency-independent) error, slightly changing the magnitude of the digitized output. Owing to the repetitive input time-multiplexing scheme, these fixed errors repeat themselves in a cyclic fashion and therefore the effect of that is equivalent to applying amplitude modulation with an M periodic sequence to the input signal series $\tilde{V}_{\text{in}}$. This introduces spurs with power that depends on the input signal amplitude and on the amount of gain error mismatch, placed at frequencies

$$f_g = \pm f_{\text{in}} + \frac{k}{M} f_s, \quad k = 1, 2, \ldots \tag{3.30}$$

- *The sub-ADC timing skew error mismatch*: similarly to the gain error mismatch, a sampling-time skew mismatch between the sub-ADCs results in an M periodic aperture sampling error in the input signal sequence. In this case, then, the effect is analogous to phase modulating (instead of amplitude modulating) the input signal with the cyclic timing skew error sequence. This time the power of the resulting spurs depends on the amount of timing skew mismatch, but also on the input signal frequency (unlike the gain error mismatch). The frequency location of these spurs is the same as that of the tones introduced by the gain error mismatch, namely

$$f_g = \pm f_{in} + \frac{k}{M} f_s, \quad k = 1, 2, \ldots \tag{3.31}$$

However, again, the power of these spurs increases with the input signal frequency since, intuitively, the faster the input signal is the larger will be the error introduced when the sampling time gets erroneously skewed.

- *The sub-ADC bandwidth error mismatch*: each sub-ADC's SHA has a certain frequency response. This frequency response has different gain and phase at different input frequencies. Mismatches between these M frequency responses convolute, again, with the digitized input sequence, leading to spurious tones at, again, the same frequencies as the other spurs. Generally, the resulting spurious energy introduced by the phase mismatches dominates over the contribution due to gain mismatches for signals at frequencies lower than the bandwidth. Conversely, the effect of gain mismatch dominates over phase mismatch for signal frequencies greater than the bandwidth.

The important thing to understand is that the introduction of the "interleaving spurs" results from the mismatch between these individual sub-ADC errors, not from the *absolute* errors themselves. In principle, if the sub-ADCs have those errors (offset, gain . . .), but such errors do match one another in all channels, then there are no interleaving spurs and only the accuracy shortcomings due to the individual channel errors are present in the output D_{out}. Clearly, if the errors are minimized in absolute terms, their mismatches are consequently smaller quantities, and therefore the spurs become smaller accordingly.

For the simplest (and most common) case of a two-way interleaved ADC, considering the first Nyquist band and a pure sinusoidal input $V_{in}(t) = A \cos(2\pi f_{in}t)$, the interleaving spurs are those located at frequencies 0 and $f_s/2$, due to offset mismatch, and the one at $f_s/2 - f_{in}$ for all other cited mismatches. The latter spur results from a combination of all of the effects discussed above. For this case, a formula combining offset, gain, and timing mismatches for the resulting interleaved output has been published in [259]:

$$D_{out}(kT_s) = A_s \cos(2\pi f_{in}kT_s + \theta_s)$$

$$+ A_n \cos\left[2\pi \left(\frac{1}{2}f_s - f_{in}\right)kT_s + \theta_n\right]$$

$$+ OS_{cm} + \cos(\pi k)OS_{diff} \tag{3.32}$$

where $G = (G_1 + G_2)/2$ is the average gain, $\mathrm{OS_{cm}} = (\mathrm{OS}_1 + \mathrm{OS}_2)/2$ is the average offset, $\mathrm{OS_{diff}} = (\mathrm{OS}_2 - \mathrm{OS}_1)/2$ is the offset mismatch,

$$A_s = AG\sqrt{\cos^2(\pi f_{in}\,\delta t) + \alpha^2\,\sin^2(\pi f_{in}\,\delta t)}$$

$$A_n = AG\sqrt{\alpha^2\,\cos^2(\pi f_{in}\,\delta t) + \sin^2(\pi f_{in}\,\delta t)} \qquad (3.33)$$

$$\theta_s = \arctan(\alpha\,\tan(\pi f_{in}\,\delta t))$$

$$\theta_n = -\arctan(\alpha\,\tan(\pi f_{in}\,\delta t)/\alpha)$$

with $\alpha = (G_2 - G_1)/(2G)$ the relative-gain mismatch and $\delta t = \delta t_2 - \delta t_1$ the timing-skew mismatch.

The first tone in Eq. (3.32) is the intended input tone. The second tone is the interleaving spur at $f_s/2 - f_{in}$ resulting from the gain and timing mismatch. The remaining terms are the spur at DC (i.e. $\mathrm{OS_{cm}}$) and at $f_s/2$ (i.e. $\cos(\pi k)\mathrm{OS_{diff}}$) resulting from offset mismatch only.

Note from Eq. (3.33) the interaction of gain and timing mismatch in the spur at $f_s/2 - f_{in}$ while there is no contribution from the offset in this term.

Indeed, that is the case assuming a pure sine signal. Harmonics will experience the same fate as that of the pure tone. Namely, they will mix with the mismatches as described above, and then lead to corresponding interleaving spurs with commensurate power.

The interleaving spurs have been thoroughly analyzed with respect to their impact on the dynamic performance of TI ADCs in tens of publications (e.g [258, 259, 260] and the several publications cited therein). The interleaving spurs are minimized either by calibrating for the above-cited mismatches [261, 262, 263] or by calibrating the individual errors to the point that also the mismatches become negligible [73, 129].

Given the parametric sensitivity of the interleaving spurs alluded to above and the fact that some of the effects of these mismatches interact with each other [260], it is not always easy to measure, separate, and correct for them. On a scale of difficulty it is possible to say that calibrating for offset mismatch is the easiest of all. There are countless ways to cancel out offset in absolute terms [33, 34, 65, 261, 262]. Besides, as mentioned above, this type of error does not depend on the input signal. The second in terms of difficulty is the calibration of gain error mismatch [261, 262, 264, 265]. Although this also depends on the amplitude of the input signal, it does not depend on its frequency.

The hardest effects to correct for are the timing skew mismatch and the bandwidth mismatch. Digital filtering (emphasis) to counter the effect of analog bandlimiting/filtering in each channel has been proposed to tackle bandwidth mismatch problems [263]. However, its implementation is far from being trivial. A common brute-force approach consists of maximizing the input bandwidth of the SHAs by boosting the sampling switches so that the bandlimiting effects are negligible at the frequencies of the input signal. Timing skews are indeed quite problematic. The brute-force approach to deal with this problem is to use a single SHA working at full sample rate f_s and then feed in

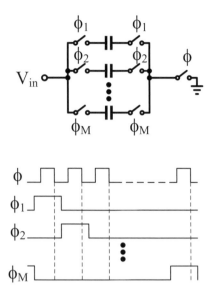

Figure 3.57 Time-interleaved passive sampling using a global clock ϕ. The falling edge determines the actual sample time for each of the M TI sampling networks [74]. © 2000 IEEE.

all the interleaved channels [266]. This approach is extremely challenging for the speed at which the SHA needs to operate and, consequently, also power-hungry.

A digital calibration approach using interpolation to correct for the proper sampling instants has been proposed in [267], and a mixed-signal approach has been proposed in [268]. However, these approaches can be computationally challenging. Moreover, what is particularly critical and challenging is the ability to properly measure the actual skews with sufficient accuracy for them to be properly calibrated. Other ways to calibrate for timing skews have relied on "blind" calibration [269, 270].

In cases where the TI ADC accuracy specifications allow it, timing skews are minimized by design using careful circuit and layout design of the sampling clock distribution network [73, 129, 265]. An alternate popular approach consists of gating the sampling time for all channels using a single (master) clock [271, 272] as shown in Fig. 3.57. Different variations of the same concept have been proposed, each one with its own advantages and disadvantages [117, 163, 273, 274]. When the number of channels M is large, a compromise between using a single SHA for all the channels and using M interleaved SHAs is to do a "hierarchical" interleaving. This means dividing the interleaved channels into groups and then using one SHA for each group while all the SHAs are time-interleaved [257, 275]. Finally, another possibility is to use digitally adjustable delays in the clock path [116, 276].

Analogously to DEM, randomizing the order of the channels to break the cyclic pattern leading to the spurious tones and spreading the associated spurious power into the noise floor has been proposed in different ways in [277, 278, 279, 280] to address all of the mismatches discussed above.

It should be noted that, although calibration and shuffling are effective ways to address the above dynamic performance issues, these also come with additional overhead, namely they add power and area to the consumption of the interleaved ADCs. Although digital functionality and power scale very well in advanced fine-line CMOS processes, this aspect should not be completely overlooked, especially when interleaving is sought for energy-efficiency reasons.

To conclude this section, we briefly mention the extremely interesting idea introduced in [163]. Rather than trying to calibrate for mismatches among the channels, the 36-channel, 5 b, 250 MSPS ADC presented in [163] integrates many more channels than it needs to interleave. This redundancy in the number of channels allows one to select the 36 "best" channels, namely those with the tighter matching, to use for the resulting TI ADC. Note that redundancy and calibration are somewhat complementary to the TI ADC matching problem. Redundancy is effective when a very large number of channels is considered since the same number of redundant channels gives greater improvement in yield than in the case of few channels. However, for high resolution and a limited number of sub-ADCs, calibration is preferable, especially since the associated overhead is minor compared with the power and area of a high-dynamic-range ADC. So redundancy works best for large arrays of medium-to-low-resolution sub-ADCs. Finally, note that the selection of the "best channels" for the prototype in [163] was performed in the foreground using off-chip software that assesses the individual performance of each of the individual channels.

3.5   What about pipelined ADCs?

The presence of pipelined ADCs has been felt throughout this chapter. For example, comparator-based/zero-crossing pipelined ADCs are discussed in Section 3.2.1, correlated level-shifting pipelines are presented in Section 3.2.2, digitally assisted pipelines are the topic of Section 3.2.4, hybrid pipelines are covered in Section 3.3, and time-interleaved pipelines are discussed in Section 3.4. Explicit or implicit references to this architecture have repeatedly been made. The fact is, the traditional pipelined ADC is a fairly mature architecture that has seen considerable, continuous, and broad development and expansion over the last 20 years. Today there is virtually no data-conversion design organization, large or small, that hasn't got traditional pipelined ADCs in its arsenal.

From a purely architectural standpoint there haven't been many major breakthroughs for traditional pipelined ADCs in very recent years. Performance has primarily advanced in terms of conversion efficiency (i.e. power consumption for a given SNDR) and sample rate. Both have progressively improved thanks to a combination of design dexterity, various circuit techniques (many of which have already been covered in the context of other architectures), and advances in process technology and calibration. In fact, during the last 5–10 years, major attention has been paid to the introduction and application of several calibration techniques for all sorts of circuit impairments: capacitor mismatch, finite op-amp gain, offsets, incomplete settling, memory effects etc. [48, 98, 99, 101,

244, 245, 246, 247, 248, 281, 282, 283, 284, 285, 286]. As has previously been discussed, both for other architectures and in a broader context, calibration is not only a way to address manufacturing limitations and parameter spread, but, if applied carefully, also can be exploited as a way to push performance. For example, mixed-signal designers know all too well that, by ignoring PVT variations and mismatch, it is possible to design circuits with minimal current consumption and minimal transistor sizes, and hence small associated parasitics, that are thus able to operate at very high frequency. The problems, so to say, arise quickly when the nominal design needs to be made robust. That is when calibration can be introduced to counter all of the cited issues with, one hopes, minimal additional loading on the signal path (hence preserving the speed and power optimization designed under nominal conditions) and at the cost of peripheral overhead. A discussion of calibration techniques for pipelined ADCs would require considerable space and is beyond the scope of this book.

Another recent trend, driven by the desire to reduce power consumption, has been associated with techniques aimed at removing an active sample-and-hold (SHA) stage [69, 287, 288, 289, 290, 291]. That is motivated by the fact that in a typical pipelined ADC the front-end SHA consumes about 20%–30% of the total ADC power and often limits the linearity and dynamic range of the ADC. Moreover, removing the SHA allows one to reduce the input sampling capacitor sizes due to the relaxed noise budget.

The removal of the SHA was first presented for low-power operation in [287]. While power consumption is reduced, there are two main drawbacks. First, for a given power dissipation, speed is limited because one requires a short comparison phase between the sampling and amplification phases of the first stage and an additional feedback capacitor decreases the feedback factor. This issue can be relaxed by dividing the sampling phase into two, using the second half of the divided phase for comparison [290]. However, this division is not suitable for a high-speed, high-resolution ADC, which demands high-linearity input sampling and a short sampling phase. Furthermore, clock jitter performance worsens because the sampling clock signal has to propagate through additional gates.

The second drawback is an aperture error caused by resistance/capacitance (RC) delay mismatch between the input networks for the first multiplying digital-to-analog converter (MDAC) and the flash ADC. This mismatch can introduce significant sampling errors, especially, when high-frequency signals are applied to the ADC [287, 288]. Thus, the input frequency can be limited by aperture error.

An approach that merges the SHA with the first MDAC by sharing on different phases both the op-amp and some of the capacitors has been proposed in [291]. This allows one to save both power and area with a technique that is a sort of extension of the well-known "op-amp sharing" technique [292, 293].

A diagram of the pipelined ADC with the merged SHA and first residue/MDAC stage (SMDAC) is shown in Fig. 3.58. A simplified schematic diagram of the SMDAC is reported in Fig. 3.59. The circuit essentially operates in three phases: sample/amplification (S/A), discharge (D) and hold phase (H). A sketch of the timing diagram

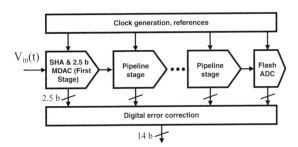

Figure 3.58 A pipelined ADC with merged SHA and MDAC [291]. © 2008 IEEE.

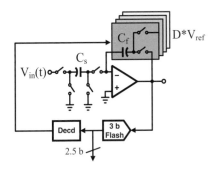

Figure 3.59 Merged SHA and MDAC circuit using op-amp and capacitor sharing [291]. © 2008 IEEE.

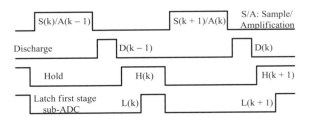

Figure 3.60 Waveforms for the control phases driving the operation of the circuit in Fig. 3.60 [291]. © 2008 IEEE.

for the corresponding digital control signals is reported in Fig. 3.60. A pictorial description of the circuit operation is shown in Fig. 3.61. The circuit operates as follows.

1. During the S/A phase, the input is sampled on the sampling capacitor (C_s).
2. While the input sample is held on C_s, the op-amp and feedback capacitors (C_fs) are re-set to an AC ground during the discharge phase.
3. Then, in the hold phase, the charge on C_s is transferred to the C_fs and the comparators make a decision after the op-amp has entered the linear-settling region.

Capacitor sizes are carefully chosen to meet noise and matching requirements, and C_s is set to $4C_f$ for a gain of 4. In the next S/A phase, C_s is charged to the next input sample,

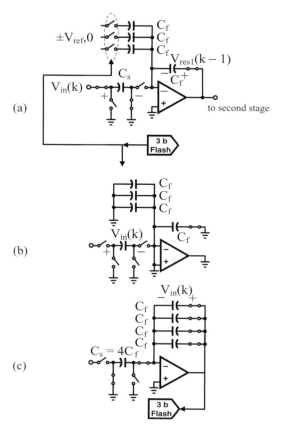

Figure 3.61 Operation of the merged SHA/MDAC circuit. (a) Sampling V_{in} at clock period k on C_s while creating residue for the sampled data of clock period $k - 1$. The latched sub-ADC drives the MDAC capacitors to the proper reference levels derived from the digitized $V_{in}(k - 1)$. (b) Holding $V_{in}(k)$ in C_s while the MDAC capacitors are discharged. (c) The hold phase: $V_{in}(k)$ develops on the MDAC capacitors by charge transfer from C_s, and it is digitized by the sub-ADC of that stage [291]. © 2008 IEEE.

while the charge on the C_fs is directly used for the MDAC operation, that is, one of the four capacitors C_f is connected to the op-amp output and the others are connected to either $\pm V_{ref}$ or 0, depending on the decision from the previous phase. The rest of the pipelined stages work with a traditional two-phase non-overlapping clock.

Some important observations need to be made. First, the hold phase is about 30% shorter than that of a conventional SHA because of the additional discharge phase. However, since the SMDAC output is not sampled by the following stage during the hold phase, the SMDAC op-amp does not see an explicit load from the sampling capacitors during the subsequent stage. Thus, the total op-amp load in the hold phase is cut by about 50% compared with that for the conventional SHA. This results in faster settling without increasing power dissipation. The second important fact relates to the memory effect. There is no memory effect for the hold operation because of the discharge phase. However, since there is no re-set phase between the hold and S/A phases, the error voltage

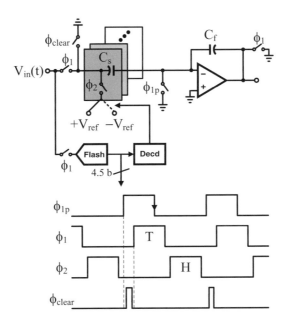

Figure 3.62 An alternate merged SHA/MDAC (SMDAC) scheme [69]. © 2009 IEEE.

stored on the op-amp input parasitics during the hold phase can affect the SMDAC output during the S/A phase. However, as analyzed and discussed in [291], this effect is actually smaller than that of a conventional MDAC.

An alternate approach for merging the SHA and the first MDAC is discussed in [69], and a simplified version of its circuit is shown in Fig. 3.62. Unlike the previous scheme, the circuit of Fig. 3.62 uses the same set of capacitors C_s both for sampling the input V_{in} during ϕ_1 and to perform the MDAC function during ϕ_2. Because of that, the charge stored in these capacitors at the end of ϕ_2 is a coarsely quantized (nonlinear) version of $V_{in}(k)$. So, when the following phase ϕ_1 starts, the circuit driving the C_ss would have to force the new input level $V_{in}(k + 1)$ against this very nonlinear initial condition and, if it does not fully settle before ϕ_{1p} ends, then nonlinear distortion would be introduced. To solve this issue an additional switch controlled by the narrow pulse ϕ_{clear} shortly clears the charge stored in the capacitors C_s before the new input is sensed.

Another difference between this approach and the previous one is that the sub-ADC (the flash ADC) in this scheme actually samples the input directly, while in the previous scheme the flash was driven by the op-amp. Because of that, the flash ADC used in Fig. 3.62 must be a "sampling flash architecture" to minimize the mismatch between the signal sampled by the MDAC and the flash while quantizing high-frequency inputs [69, 287, 288].

We'll conclude this section with a brief mention of a very interesting architectural innovation for pipelined ADCs that was presented recently in [294]. The advantages of continuous-time $\Delta\Sigma$ ADCs versus discrete-time $\Delta\Sigma$ ADCs have been discussed in

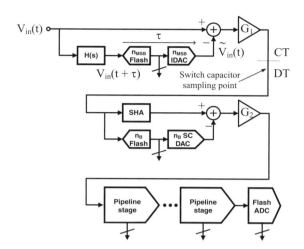

Figure 3.63 A continuous-time pipelined ADC with a first CT stage and subsequent DT stages [294]. © 2010 IEEE.

detail in Section 3.1.4 and include inherent anti-aliasing filtering, ease of drivability, and different sampling jitter sensitivity among others. Similarly, [294] presents two different pipelined ADC architectures with a continuous-time front stage that have features similar to those of CT $\Delta\Sigma$ ADCs. In fact, the first stage of a pipelined ADC is being implemented using a continuous-time circuit whose function is to synthesize the first residue and to resolve the first n_{MSB} most significant bits. This residue is subsequently fed to a traditional switched capacitor pipelined ADC which samples it and completes the conversion in the usual discrete-time fashion. Just like in CT $\Delta\Sigma$ ADCs, since the sampling does not happen at the input of the ADC but rather after some CT gain stage, the input signal has inherently been pre-filtered for aliasing,[23] and, furthermore, the corresponding sampling jitter sensitivity of this sampling is relaxed by the amount of this intermediate gain. On the other hand, the jitter sensitivity can emerge in a different guise when the quantized input is compared/subtracted with/from the continuous input signal to synthesize the residue.

A diagram depicting this architecture is shown in Fig. 3.63. On comparing its CT first stage with a classic switched capacitor's DT second stage it is possible to see that the first stage is aiming to derive the first residue using CT circuitry. However, the cascade of the sub-flash and DAC requires some time τ to produce the analog quantized input $\hat{V}_{in}$ which, subtracted from the continuous-value input V_{in}, creates the residue. In the classic switch capacitor circuitry (e.g. the one for the second and following stages) this issue is circumvented by the fact that the input V_{in} is sampled before its quantized $\hat{V}_{in}$ analog is synthesized and then subtracted from it to form the residue. Therefore both V_{in} and $\hat{V}_{in}$ correspond to the very same original voltage level and same time instant. In the first stage of the structure of Fig. 3.63, instead, it takes at least τ to quantize V_{in} and then synthesize $\hat{V}_{in}$. Therefore, in order to have the subtraction with the proper timing and

---

[23] Although in CT $\Delta\Sigma$s the oversampling dramatically facilitates the anti-alias filtering since it allows a relaxed roll-off and allows sufficient frequency range to achieve significant attenuation at frequencies approaching $f_s/2$.

proper voltage level, a prediction filter $H(s)$ is introduced in front of the quantizer/DAC block. The function of the prediction filter is to create an estimate of the input V_{in} at a time τ seconds later, so that this signal, after subsequently having been delayed by τ seconds by the quantizer/DAC block, ultimately results in the proper $\hat{V}_{in}(t)$ when it is used to create the residue.

Although the heralded advantages of a CT front end are certainly very attractive, it is quite apparent from this discussion that the design of the prediction filter $H(s)$ requires one to make some assumptions and impose restrictions on the frequency content and allocation of the input signal. Furthermore, the predicted signal is, itself, an approximation, which sets limits on the accuracy of the converter, and, finally, the circuit implementation of $H(s)$ could be non-trivial.

3.6   Conclusions

This chapter has covered some of the most recent advances in analog-to-digital conversions. Dynamic performance and energy efficiency have progressed considerably during the last decade or so. The increasing challenges offered by nanometer-scale CMOS processes have sent designers back to the proverbial white board, and have led to a wide variety of different solutions for the many different specifications and needs in ADC design. Some of these are purely analog, while others rely on digital functionality to various degrees. The proper balance depends on the context into which the ADC needs to fit and, to some extent, also on the designer's personal judgment and experience.

4 Advanced digital-to-analog converters

As briefly mentioned in Section 2.4, a variety of DAC architectures have been developed and are being utilized, depending on signal-bandwidth and dynamic-range needs. These include capacitive DACs, resistive DACs, and current-steering DACs among others. The former two categories are routinely used either for stand-alone converters for precision applications or as internal sub-blocks for all types of ADCs. However, the current-steering architecture is certainly the one which has recently received the most attention, across a large resolution range, due to its inherent high speed and its ability to directly drive low-impedance loads (notably resistors). It is for this reason that this chapter will cover only current-steering DACs.

The chapter is organized as follows. First the basics of this architecture are briefly summarized for convenience. That includes a discussion of segmentation. Static linearity is then the topic of Section 4.2, where the causes of static linearity errors and techniques by which to minimize them are discussed. That is followed by a discussion on "intrinsic-matching DACs" where static linearity is guaranteed by design. Next, traditional as well as advanced calibration techniques are covered as an alternate approach to address mismatch errors, freeing up the designer to focus on other dimensions such as silicon area or dynamic performance.

Various sources of dynamic linearity errors are discussed in Section 4.3. Their detrimental effects leading to high-frequency distortion are described. This is followed by an account of a variety of circuits and techniques to mitigate these issues in Section 4.3.3.

Layout and floorplanning techniques are particularly important to the performance of current-steering DACs since improper layouts cause as many real-life problems as (or, arguably, even more than) poor circuit design. This is the topic of Section 4.4.

Subsequently, dynamic element matching (DEM) for Nyquist DACs is discussed in Section 4.5. These techniques are used to randomize mismatches so that the associated spurious tones arising from harmonic distortion are turned into broadband noise-like spurious content, trading off an improvement in SFDR for a degradation in NSD.

Signal processing techniques, which are increasingly being implemented in the context of digital-to-analog conversion, are discussed in Section 4.6. These include interpolation and modulation of the output pulse to shape the classic $\text{sinc}(x)$ output spectral distortion and/or to modify the output images in bands higher than Nyquist, etc., which are covered in Section 4.6.

A brief discussion of "specialty-type" DACs such as direct digital synthesis (DDS) and audio DACs as well as the rapidly emerging area of so-called RF-DACs is the topic of Section 4.7.

The topic of current-steering DACs is much broader than many non-specialists would guess at first, and, therefore, a lot has had to be left out of this book due to space limitations. However, perhaps, one of the messages that the reader will take away from carefully reading this chapter is that the (too) common misconception that little is happening in the "land of DACs" (or the even more puzzling one that "designing DACs is easier than designing ADCs") could not be any more baseless.

4.1   Current-steering DACs: basic architecture and segmentation

Before covering some of the most recent advances in the area of current-steering DACs a summary of the background concepts for this architecture will be given. Some people refer to current-steering DACs also as *switched current DACs*, given that currents are switched toward different nodes in order to synthesize an analog output. Although we won't digress into a semantic debate and in spite of the fact that some *current copier* cells sometimes find use in the internal circuitry of current-steering DACs [295], we will not use the term "switched current" when referring to these DACs because that term refers to a fairly different class of analog circuits meant to be the current-mode alternative to switched-capacitor circuits [296].

For the sake of completeness we will also mention that the name "continuous-time" (CT) DACs is also sometimes used for current-steering DACs simply to distinguish them from switched-capacitor (SC) DACs (which some would find natural to associate with discrete-time systems) [154].

A current-steering DAC converts an n-bit input binary code of the form $\mathbf{D}_{\text{in}} = (d_{n-1}, \dots, d_1, d_0)'$ (where $d_i = 0$ or 1 are the input bits) into an output current I_{out}:

$$I_{\text{out}} = \frac{I_{\text{FS}}}{2^n - 1} \cdot \sum_{i=0}^{n-1} d_i 2^i \tag{4.1}$$

where I_{FS} is the full-scale output current.

Equation (4.1) can be rearranged as follows:

$$I_{\text{out}} = \sum_{i=0}^{n-1} d_i \frac{I_{\text{FS}} 2^i}{2^n - 1} = \sum_{i=0}^{n-1} d_i I_i \tag{4.2}$$

which suggests that, having n binary weighted currents $I_i = I_{\text{FS}} 2^i / (2^n - 1)$ available, I_{out} could be obtained by properly steering all the currents I_i corresponding to the bits $d_i = 1$ toward the output node (where they get summed), while all the remaining currents I_j for the bits $d_j = 0$ would be steered toward a ground/dump node, as depicted in Fig. 4.1. This possible implementation is referred to as a *binary weighted* DAC.

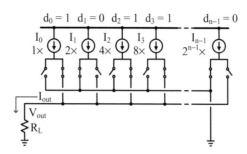

Figure 4.1 An ideal binary weighted DAC.

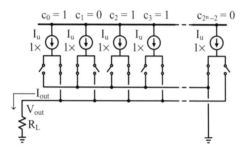

Figure 4.2 An ideal thermometer decoded DAC.

On the other hand, if $\mathbf{C} = (c_{2^n-2}, \ldots, c_1, c_0)'$ is the thermometer coded equivalent of $\mathbf{D_{in}} = (d_{n-1}, \ldots, d_1, d_0)'$, then Eq. (4.1) can also be rearranged as

$$I_{out} = \sum_{i=0}^{n-1} d_i \frac{I_{FS} 2^i}{2^n - 1} = \sum_{j=0}^{2^n-2} c_j \frac{I_{FS}}{2^n} = \sum_{j=0}^{2^n-2} c_j I_u \qquad (4.3)$$

which suggests that, having $2^n - 1$ identical "unit" currents $I_u = I_{FS}/2^n$ (sometimes referred to as *unary* currents), another possible implementation would consist of steering all the currents I_u corresponding to $c_j = 1$ toward the output node synthesizing I_{out}, while, again, steering all the remaining unary currents to the dump/ground node, as shown in Fig. 4.2. This implements a *thermometer decoded* DAC.

Mixed binary-weighted/thermometer-decoded implementations can then be conceived. For example, if the M most significant bits of $\mathbf{D_{in}}$ are thermometer decoded while the remaining L least significant bits of $\mathbf{D_{in}}$ are left in binary code, then the summation in Eq. (4.1) can be partitioned into two summations, one with a binary weighted contribution and one with a unary contribution:

$$I_{out} = \frac{I_{FS}}{2^n - 1} \cdot \sum_{i=0}^{n-1} d_i 2^i = \sum_{i=0}^{L-1} d_i I_i + \sum_{j=0}^{2^M-2} c_j I_u \quad \text{with } n = M + L \qquad (4.4)$$

This leads to the implementation of a *segmented* DAC, made of two sub-DACs: a thermometer decoded *MSB segment* and a binary weighted *LSB segment* as shown in Fig. 4.3.

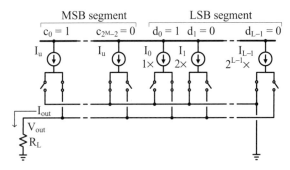

Figure 4.3 An ideal segmented DAC.

Note that in Eq. (4.4), to satisfy the equality, the scale of the two sub-DACs must be properly re-sized, leading to the following unary and weighted currents:

$$I_u = \frac{I_{FS}}{2^M}$$

$$I_i = \frac{I_u 2^i}{2^L}$$

(4.5)

and to the following additional relationships:

$$I_u = I_0 + \sum_{i=0}^{L-1} d_i I_i$$

(4.6)

$$I_0 = \frac{I_{FS}}{2^n}$$

which essentially means that the full-scale current I_{FS} is divided into the 2^M unary currents (Eq. (4.5)); $2^M - 1$ of them are used to implement the MSB segment while the last one can be further sub-divided in a binary weighted fashion to implement the LSB segment (Eq. (4.6)).

The LSB segment just defined can actually be segmented further (in a somewhat recursive fashion if we think of this LSB segment as an independent DAC that we now want to segment more) by thermometer decoding its U most significant bits while leaving the remaining least significant bits implemented as a binary weighted DAC. So the n-bit DAC is an "M–U–L segmented DAC" composed of an M-bit thermometer decoded (unary) MSB segment, a U-bit thermometer decoded *upper*[1] LSB segment and, finally, an L-bit binary weighted *lower* LSB segment [297, 298].

It is apparent, at this point, that there are many options with regard to the means of synthesizing the output current I_{out} of a DAC by steering and summing, in an (input) code-dependent way, a set of pre-built unary and binary weighted currents. But before analyzing advantages, trade-offs, and motivations behind the above architectural choices, let us begin to look at some circuit implementations to fix the ideas.

---

[1] Sometimes referred to as an "intermediate" segment.

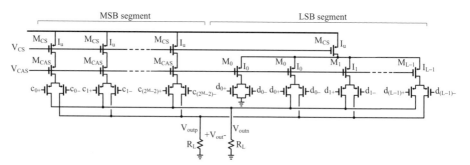

Figure 4.4 Simplified transistor-level implementation of a segmented DAC.

A possible transistor-level implementation for the M–L segmented DAC of Fig. 4.3 is shown in Fig. 4.4. The MSB and the LSB segments are visible on the left- and right-hand side of the diagram, respectively, following the structure previously shown in Fig. 4.3. The unit currents I_u are set by the 2^M identical PMOS devices M_{CS}, all identically biased with the same source, gate, drain, and bulk potentials. As previously described, the first $2^M - 1$ PMOS M_{CS} devices are used in the MSB segment. The last M_{CS} device on the right sets the total current used in the LSB segment.

This current is then partitioned into the binary weighted derivative currents I_i using a current splitter constituted by cascode devices M_0, M_1, ... For the latter devices, it is assumed too that their corresponding terminals' potentials are all equal and, in fact, identical to those of the cascode transistors M_{CAS} of the MSB segment. Hence the proper current split is obtained through device sizing (i.e. suitably choosing the W and L of all the cascodes). It is assumed that all of the above-mentioned devices, so far, are always biased in saturation.

Continuing from top to bottom, all cascode devices are followed by source-coupled pairs acting as *current-steering switches*. The gates of these switches are driven by switch driver circuits (not shown in the picture to avoid clutter) so that each current supplied by the cascodes is fully steered (through the turned-on switch) toward one of the two output nodes at potentials V_{outp} and V_{outn}, respectively. Note, in fact, that in this transistor implementation, unlike in the conceptual schemes of the prior figures, a differential output with a ground-referenced load R_L is implemented (giving a differential output voltage $V_{out} = V_{outp} - V_{outn}$ and differential output current $I_{out} = I_{outp} - I_{outn}$).

Before proceeding any further, a few remarks are in order. First of all, any deviation from the above-cited assumption of saturation and equal quiescent point for the current-source devices M_{CS} will result in considerable current errors, leading to static linearity errors (INL/DNL errors). The drain of the M_{CS} devices is therefore a relatively sensitive node, and the potential at these nodes depends on the cascode devices M_{CAS}, M_0, M_1, etc. It is, in fact, equally important that the cascode devices act to keep that node as firm as possible and isolated from the rest of the structure, particularly from any code-dependent disturbance originating from the switches or, even worse, any potential variation resulting from V_{out}.

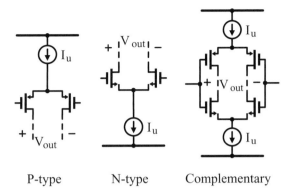

P-type   N-type   Complementary

Figure 4.5 Possible architectural styles: P-type or current sourcing (left); N-type or current sinking (center); and complementary or push/pull (right).

Second, the use of the current splitter based on the cascode devices M_0, M_1, etc. is a powerful segmentation technique, because constraining the total current of the LSB segment to be equal to I_u means that the smaller weighted currents I_i are now expected to match each other in the desired weighted ratio only to a level that is commensurate with the number of bits in the LSB segment alone. Satisfying that will automatically meet the required, but much greater, ratio between such currents and the MSB unit current I_u. For example, let us consider an $n = 14$-bit DAC with an MSB segment of $M = 7$ bits and an LSB segment of $L = 7$ bits. If the LSB segment were to be implemented using a separate array of seven weighted currents (plus an additional I_0 as explained before) and hence the value of such currents were set solely by the corresponding current-source devices, then the smallest current I_0 (which is $I_0 = I_u/2^L = I_u/128$) would need to match I_u to better than $1/2^n = 1/16\,384 \simeq 0.006\%$ (referred to, in the technical jargon, as "to $n = 14$-bit level"). Conversely, using the above current splitter, the requirement for I_0 is now only to match the other currents within the LSB array. So, considering the largest of them, I_6 (which is $I_6 = I_0 \cdot 2^{L-1} = 64I_0$), I_0 will need to match I_6 to better than $1/2^L = 1/128 \simeq 0.8\%$ (referred to as "to $n = 7$-bit level"). The tough matching challenge is then limited to making the 128 unity currents I_u properly match one another.

Note that so far we have shown DACs sourcing a current from a positive supply into a grounded load. Because the current sources are implemented using a stack of PMOS transistors, this type of architectural choice goes by the name of P-type DAC. That is shown on the left in Fig. 4.5. It is clear that a complementary approach, an N-type DAC, is also possible, as shown in the center of Fig. 4.5, where the load is tied to a positive supply and the DAC sinks its output current from it. Finally, it is also possible to have fully complementary DACs composed of a combination of a P- and an N-type DAC alternately sinking and sourcing currents from/to a load at an intermediate voltage level [299] as shown on the right of Fig. 4.5. Although, traditionally, almost exclusively P-type DACs have been designed in the past,[2] N-type DACs are becoming much more common

---

[2] Owing to the matching and $1/f$ noise characteristics of PMOS current sources as well as to make it straightforward to directly drive a ground-referred output load.

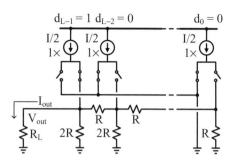

Figure 4.6 A binary weighted DAC using identical currents and an R–$2R$ network.

in very-high-speed applications due to the many performance advantages in having high-mobility (μ_n) devices in the signal path. In this chapter we will liberally switch back and forth between P-type and N-type DACs in our discussion, always implying that the complementary type (including the fully complementary type) is easily derivable by the reader.

Before concluding this section it is worth mentioning one more technique that allows relaxation of the current-source matching requirements for implementing the binary weighted LSB segment. That is based on an application of so-called "R–$2R$" ladder networks, which are more commonly utilized in voltage-mode low-speed/precision DACs [5, 52]. A binary weighted current-steering DAC using identical current sources combined with an R–$2R$ network to implement the binary weighted current contributions is shown in Fig. 4.6.

Owing to its regularity and modularity, this approach allows one to design very compact and area-efficient DACs. However, one of the main disadvantages is that the resistive network, with its many intermediate nodes (the current source's injection points) and the output node, introduces just as many time constants [52, 300]. This leads to large glitches in the output when the bits change state and, therefore, to inferior dynamic linearity. For this reason, this technique is generally limited to implementing only a few bits and it finds more frequent use in bipolar designs, such as in [301, 302, 303, 304], where many binary weighted BJTs are harder to lay out and design than their CMOS counterparts. Otherwise it can be appropriate where very-high-frequency dynamic performance is not required but small area is important, such as in calibration DACs [305] and in some video applications [300].

4.2   Static linearity

4.2.1   A summary of the limitations

The final example of Section 4.1 has already introduced a key point at the foundation of the linearity performance of DACs, namely the matching of the currents in the array. Let us compare the two extreme cases, namely a fully thermometric DAC versus a fully binary weighted DAC. As discussed earlier, the foundation of the thermometric DAC is an array of ideally identical $2^n - 1$ unity currents I_u. If a DAC based on this array is given a digital input sequence $\mathbf{D}_{in}$ corresponding to an increasing ramp with unit steps (e.g. the sequence 0, 1, 2, ...) then the resulting current output will be a monotonically increasing ramp since a new I_u current is steered toward the output at each code increment. The currents in the array will have relative mismatch to one another, but, since some current is always added at each code increment, monotonicity is preserved. However, deviations of the actual unary sources from the nominal value I_u will lead to errors in each step size and hence to DNL errors. Accumulation of such errors leads to INL errors. Deviations from I_u are mismatches of each element of this array with respect to I_u and, therefore, also mutual mismatches between the currents of the array. Note, however, that since this DAC is fully thermometric each current I_u in this example corresponds to an LSB $(I_u = I_{FS}/2^n)$ and, therefore, the relative matching of the current sources must be worse than $\pm 50\%$ of I_u to lead to a local DNL error greater than 1/2 LSB. On the other hand, it is intuitive that the accumulation of even such small DNL errors can lead to large (many LSBs) INL errors.

Providing the same digital ramp to a fully binary weighted DAC won't necessarily lead to the same smoothly monotonic output. In this case, mismatches between the weighted current sources will lead to DNL errors at major bit transitions. For example, let us consider an $n = 4$-bit DAC and imagine that the input ramp is nearing middle scale, going from $\mathbf{D}_{in} = (0111)'$ to $\mathbf{D}_{in} = (1000)'$ (namely the MSB is changing state). For $\mathbf{D}_{in} = (0111)'$ the currents I_0, I_1, and I_2 are steered toward the positive output while I_3 is steered toward the negative output. When the next code $\mathbf{D}_{in} = (1000)'$ arrives I_3 is steered to the positive output and all others to the negative output. The output is expected to increase by one LSB, and it will increase by $\Delta I_{out} = I_3 - (I_0 + I_1 + I_2)$. The deviation of ΔI_{out} from one LSB could be quite large and, in fact, it could equally well be positive or negative (leading to a non-monotonicity in the latter case) because of a mismatch in I_3 with respect to its nominal value $I_{FS}/2$. For example, a $\pm 50\%$ error on I_3 will here lead to a gigantic "middle-scale" error equal to $\pm 25\%$ of the full-scale current! Similarly, a $\pm 50\%$ error on I_2 will lead to an error of $\pm 12.5\%$ of the full-scale current at the bit transitions $\mathbf{D}_{in} = (0011)' \rightarrow (0100)'$ and also at $\mathbf{D}_{in} = (1011)' \rightarrow (1100)'$, namely at the two "quarter-scale" points. Similar considerations apply for I_1s matching at the 4-bit transitions at each eighth of the full scale and so on in a "self-similar" fashion.

Let us also note that, although, to make it easier to achieve reasonable matching, binary weighted currents can be implemented by aggregating (summing) unary currents, as exemplified in Fig. 4.7, the above-described problem at bit transitions continues to hold for binary weighted DACs.[3]

It is clear, then, that a binary weighted DAC suffers from greater DNL errors than those of a thermometric DAC. However, assuming that the mismatch errors between the unit elements used to implement either type of architecture are uncorrelated, when these

---

[3] Let us not forget that, again, at a bit transition, the output does not change by one LSB because of the addition of a single LSB current as happens in a thermometric DAC. Even using the aggregate unity-currents implementation, the output is changing because a large bank of cumulatively mismatched currents is replaced by a completely different set of mismatched currents. While the total number of unity currents involved has changed by one unit only, however, the cumulative error of the first bank can be much larger than the unit current itself, and the same is also true for the cumulative error of the new bank of current sources now steered to the output.

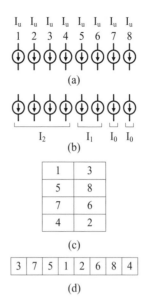

Figure 4.7 A 3-bit binary weighted current array obtained by aggregating unity currents: (a) the unity-currents array; (b) the binary weighted currents array; (c) a possible layout arrangement for the eight unit currents; and (d) an alternate layout arrangement for the same system.

individual mismatch errors quadratically add together leading to the INL, it is possible to see that, in statistical terms, the two architectures have essentially very similar INL performance [306].[4]

Insofar as DNL is concerned, a segmented DAC will perform somewhere in between the case of the binary DAC and the case of the thermometric DAC (while the INL performance, once again, is the same as for the other architectures). Also, it should be apparent now why, as shown in the previous section, the MSB segment is thermometer decoded and the LSB segment is binary weighted (not the other way around!). So for an M–L segmented DAC the worst-case DNL error is then more likely to happen at a code transition when the LSB segment turns off and a new unity current from the MSB segment turns on.

Denoting with σ_u the rms error of the unity currents aggregated to implement the above-discussed architectures, it is possible to estimate the DNL and INL errors for each choice. The corresponding results are reported in Table 4.1 [54].

Another source of static linearity problems is the finite output impedance of the current sources in the DAC array. The issue is easily understood by considering the following examples. Let us consider the thermometer decoded DAC shown in Fig. 4.8, where the impedances $Z_u = 1/(1/R_u + j\omega C_u)$ of the unary current sources have been highlighted and all sources and impedances are assumed to be identical.

---

[4] A recent study has demonstrated that, in fact, on average, for the same resolution and unity-currents accuracy, binary DACs are expected to have smaller INL errors than thermometric DACs [307]. However, in practice, these differences are small when typically high yield is targeted.

Table 4.1. A summary of INL and DNL errors for various architectural choices [54]

	Thermometric	Segmented	Binary
σ_{INL}		$\frac{1}{2}\sigma_u\sqrt{2^n}$	
σ_{DNL}	$\simeq\sigma_u$	$\simeq\sigma_u\sqrt{2^{L+1}-1}$	$\simeq\sigma_u\sqrt{2^n-1}$
Number of switched elements	2^n-1	$L+2^M-1$	n

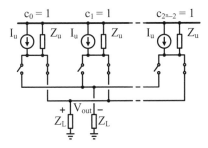

Figure 4.8 A thermometer decoded DAC with finite source impedances.

The DAC's output current is fed to a differential load Z_L, developing a differential output voltage V_{out}. If the current sources were ideal, and hence $|Z_u|$ were infinite, the output would follow from the ideal equation (4.3). However, with finite $|Z_u|$, as the input code changes and the individual unary currents are steered toward one output terminal or the other, the overall output load on each of the two output terminals varies as well. For example, when the input code is zero, the total impedance at the positive output node is Z_L while the total impedance at the negative output node is the parallel between Z_L and *all* of the Z_us (i.e. $Z_u/(2^n-1)$). When the input code is at full scale (i.e. 2^n-1) then the situation is completely reversed, namely, now it is the positive output node which sees the smallest total impedance $Z_L//[Z_u/(2^n-1)]$. On the other hand, if the input code is close to mid-scale, then half of all the Z_u are parallel to Z_L at the positive output node while the other half are parallel to Z_L at the negative output node. So, in this case the output load is essentially balanced.

Clearly, this code-dependent output impedance introduces a code-dependent error into the current-to-voltage conversion. This error is present for all frequencies, so it introduces both static and dynamic linearity problems. In fact, while at DC the error can be relatively minimal since all the impedances are (generally large and) purely resistive, at higher frequency the capacitive components of Z_u instead become increasingly dominant (so $|Z_u|$ becomes smaller and smaller), making the overall output impedance very much code-dependent.

In [308] it is shown that the static linearity is affected as follows. If the output is taken differentially, as in the example of Fig. 4.8, then the resulting INL has an S-shape. Hence, at low frequency, odd-harmonic distortion is to be expected. The corresponding maximum contribution to the INL has been estimated to be [52, 308]

$$\text{INL} \simeq \frac{R_L 2^{2n}}{4R_u} \tag{4.7}$$

However, if the output is taken single-ended (namely on only one of the two terminals, as in the case of Fig. 4.2) then the corresponding INL has a bow shape (hence leading to even-harmonic distortion). Furthermore, in the latter case, the INL is about one order of magnitude larger than that in the differential case.[5] Clearly, the linearity error due to finite current source impedance adds on top of the linearity error introduced by mismatches discussed previously (which is a completely independent issue). Suitable use of cascoding in the current sources is a way to keep Z_u high enough, with respect to Z_L, to make this effect sufficiently negligible for the desired level of linearity.

Unfortunately, as will be discussed further in Section 4.3, even with cascoding, ultimately, at very high frequency, the parasitic capacitances make the Z_us non-negligible and distortion is unavoidable. It goes without saying that, although we have presented this problem in the simplified case of a thermometric DAC, the problem very much exists for any type of segmentation.

4.2.2   Intrinsic-matching DACs

The previous section has highlighted how mismatches in the current sources or the current splitters lead to static linearity errors quantified by DNL/INL. Clearly these mismatches arise from the mismatch in the transistors setting these currents. So, in the example of the architecture shown in Fig. 4.4, mismatches in the M_{CS} devices introduce errors in the MSB segment currents, whereas mismatches in the M_0–M_{L-1} devices introduce errors in the current splitter and hence in the LSB segment. Assuming that all sources of systematic mismatch[6] in such devices have been resolved, random variations arising from the manufacturing process affect matching. For MOS devices, biased in saturation, with a relatively long and wide geometry (which, for today's standards, means $L, W > 1\,\mu\text{m}$) and roughly square shape ($W \sim L$), the relative mismatch can be quantified using Pelgrom's model [42, 43], which has already been mentioned in Section 1.2.1:

$$\sigma^2(\Delta V_T) = \frac{A_{VT}^2}{W \cdot L} \tag{4.8}$$

$$\left(\frac{\sigma(\Delta\beta)}{\beta}\right)^2 = \frac{A_\beta^2}{W \cdot L} \tag{4.9}$$

where A_{VT} and A_β are the Pelgrom parameters quantifying the expected mismatch in threshold voltage V_T and current factor β for a given process. Performing a mismatch analysis (similar to a noise analysis where random mismatches are treated analogously to random noise sources [43]) and assuming a square law for an MOS's drain current lead to the relative mismatch in drain current between two nominally identical adjacent devices with equal gate–drain–bulk–source potentials but randomly mismatched

---

[5] Even when the output is taken differentially, a small fraction of this considerably larger even-order nonlinearity will add to the differential odd-order nonlinearity as a result of small output imbalances/mismatches.

[6] Including differences, between otherwise matched devices, in the corresponding gate–drain–source–bulk potentials as well as differences in the layouts or their surrounding layers and structures.

drain current:

$$\left(\frac{\sigma(\Delta I_{DS})}{I_{DS}}\right)^2 = \left(\frac{\sigma(\Delta\beta)}{\beta}\right)^2 + \left(\frac{g_m}{I_{DS}}\right)^2 \sigma^2(\Delta V_T) \qquad (4.10)$$

or, otherwise, referring the mismatch to the gate–source voltage instead of the drain current,

$$\sigma^2(\Delta V_{GS}) = \sigma^2(\Delta V_T) + \frac{1}{(g_m/I_{DS})^2}\left(\frac{\sigma(\Delta\beta)}{\beta}\right)^2 \qquad (4.11)$$

In most practical cases, the threshold-voltage mismatch (ΔV_T) is the dominant source of mismatch, rather than the geometry/lithography mismatch ($\Delta\beta$). Its contribution can be diminished (by brute force) by designing for the largest possible overdrive voltage ($V_{GT} = V_{GS} - V_T$, typically with $V_{GT} \gg 200$–$300\,\mathrm{mV}$) and/or for the lowest possible transconductance efficiency (i.e. by making g_m/I_{DS} minimal in Eq. (4.10)). This, however, often leads to large values of V_{DS} (since the devices need to be biased in saturation) and hence comes at the expense of voltage headroom, which is an especially critical issue in scaled CMOS processes with low supplies.

Of course, as can be seen in Eqs. (4.8) and (4.9), increasing the MOS area $W \cdot L$ decreases both sources of mismatch; but large areas also clearly lead to large parasitics that impact the high-frequency performance of the DAC.

The previous equations have given us a model for the random mismatches in MOS devices. Static linearity specifications can then be addressed either by designing for "intrinsic matching" or by calibration. Designing for "intrinsic matching" means that the mismatch level of the critical currents is deliberately kept sufficiently low by design, namely by a suitable combination of segmentation choices (i.e. how many bits are assigned to the MSB, U-LSB, and L-LSB segments and how these are implemented), device sizing, and biasing. In cases like these, designers can proceed by using/iterating "Monte Carlo"-style simulations to determine the best choice of segmentation, device sizing, and biasing [309, 310]. For instance, in its simplest form, Pelgrom's model can be used to randomly generate "populations" of mismatched current sources. These randomly generated current sources are then suitably combined (e.g. using the aggregation technique discussed in the previous section) into simulated segmented DACs (each one with its own DNL and INL), ultimately allowing one to analyze distributions of INL/DNL and, hence, estimate the yield for a desired INL for given choices of device sizing, biasing, and segmentation. This type of analysis can be done fairly quickly using tools such as Matlab if, just to fix the ideas, a sample DAC (one member of a large population of simulated DACs) can be simply internally represented as a one-dimensional vector whose elements are the nominal values of its current sources with their added random mismatch obtained from Eqs. (4.8) and (4.9). This Monte Carlo analysis (generating such DAC populations) is then iterated, varying segmentation and device design, until a satisfactory linearity and corresponding yield are reached.

Alternatively, mathematical models for the yield and distribution of DACs' INL have been proposed. A recent one, derived from the "Brownian bridge" process, gives the

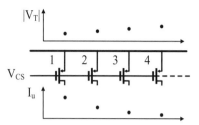

Figure 4.9 A linear gradient in V_T over a row of identical devices results in a large mismatch in drain current (further aggravated by the current square law) between distant devices.

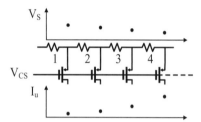

Figure 4.10 An appreciable "$r \cdot i$" drop over a supply line of a row of identical devices results in a large mismatch in drain current between distant devices.

yield for a case with $|INL_{max}| < 0.5\,\text{LSB}$ as [307]

$$\text{Yield}_{\text{INL}} = 1 - 2\sum_{k=1}^{\infty}(-1)^{k-1}\exp\left(-\frac{k^2}{2(2^n - 1)(\sigma(\Delta I_{DS})/I_{DS})^2}\right) \quad (4.12)$$

allowing one, again, to determine the target relative current mismatch ($\sigma(\Delta I_{DS})/I_{DS}$) for a desired yield.

The previously discussed random mismatches are the dominant source of matching errors on a very localized scale, namely when comparing adjacent devices. However, superimposed on this local randomness are process and, possibly, temperature gradients, namely variations of process/electrical parameters in space (over the die) leading to mismatch between nominally identical devices placed at different locations the chip. For example, as shown in Fig. 4.9, a linear gradient in threshold voltage V_T over a row of identical transistors leads to a large mismatch in drain current between distant devices. A similar effect is observed when, for example, due to ohmic (also known as "$r \cdot i$") voltage drops over the supply line of the same structure as that of Fig. 4.9, the source potentials of the transistors are gradually shifted from their nominal voltage over the array, leading to appreciable spatial mismatch in the currents as shown in Fig. 4.10.

If, for instance, the devices in one of the previous two examples set the currents for a thermometric DAC where an increasing input code corresponds to steering the next current in this row, then both the DNL and, especially, the INL will rapidly increase over the input code. Again, since the order in which the currents are steered, called the

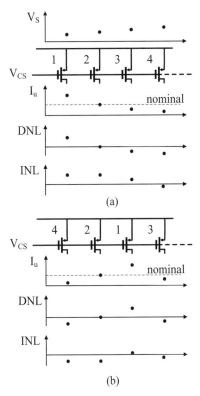

Figure 4.11 The effect of a gradient to the DNL and INL in the cases of (a) a straight switching sequence and (b) a better switching sequence that minimizes error accumulation in the INL.

"switching sequence," is the same as the order in which these are placed in space, spatial errors directly result in increasing static linearity errors.

Since gradients cannot be eliminated, one way to mitigate this problem is to adopt a cleverer switching sequence. Let us consider again the example of the threshold voltage gradient shown in Fig. 4.11(a). If, instead of following the depicted switching sequence "1, 2, 3, 4," the switching sequence is changed to "4, 2, 1, 3," as shown in Fig. 4.11(b), then the DNL sequence is changed (though its amplitude is not changed) into an alternating positive/negative-sign sequence and, therefore, the INL errors now compound differently into a much more bounded amplitude set.

Clearly, as we have already remarked, the error build-up from other types of spatially induced mismatches, such as ohmic drops on power lines or undesired propagation delays, can also be mitigated through the use of appropriate switching sequences.[7] Furthermore, when considering large spatial arrangements, gradients might no longer be that linear and may have quite arbitrary contours.

---

[7] It is important to note, however, that $r \cdot i$ drops or propagation delays, unlike a-priori unknown gradients, are predictable in size and distribution and, therefore, can also be dealt with by more effective techniques, such as equalization by means of layout trees [311].

In [312, 313, 314] optimal switching sequences to tackle arbitrary gradients on both one-dimensional and two-dimensional arrays have been discussed. Fifty-fold reductions in INL have been claimed to have been achieved through sensible combination of unit-element splitting/distribution together with switching-sequence optimization using the so-called "Q^2 random-walk" switching scheme [312]. A simplified approach to element placement and switching-sequence design that does not require quad and multiple-quad decompositions is described in [314].

In most Nyquist-rate DACs, with a relatively large number of current sources, the switching sequence will be fixed once. This means that it will be "hardwired" to the DAC as part of its design. On the other hand, if the number of current sources involved is not too high (say, for example, fewer than 20 or so are used), and if the DAC is oversampled (as are those used in the feedback DACs of oversampled $\Delta\Sigma$ ADCs) then it is practical to design circuits that change the switching sequence for each new data input to the DAC. The latter approach is termed "dynamic element matching" (DEM) [67, 68, 81]. By constantly changing the association between the DAC's input code and the mismatched elements used to synthesize the output, DEM algorithms attempt to de-correlate mismatches from inputs and, hence, to turn mismatch-induced linearity errors into noise-like code-independent spurious outputs. Dynamic element matching will be discussed in Section 4.5.

4.2.3   Calibration of DACs

The main advantage of intrinsic-matching DACs is their relatively low complexity. The main disadvantage, however, is that adopting intrinsic matching dramatically reduces the design degrees of freedom since the quiescent point and sizing of essentially all current sources and splitters become fixed to meet the matching requirements. Unfortunately, as pointed out in the previous sections, this can lead to designs with either large associated parasitics or limited voltage headroom, or both, with adverse consequences for the high-frequency performance of these DACs.

Calibration is then an alternative to intrinsic matching. In other words, sizing and biasing are not being bounded by matching constraints because the effects of mismatches in the currents are measured and then compensated for in a different manner.

The simplest form of analog calibration of DACs' current sources is based on the current-copier technique [295]. The principle of that is depicted in Fig. 4.12 [296].

First, an input current I_{in} is fed into a diode-connected MOS transistor biased by I_b, hence developing a corresponding gate–source voltage $V_{GS}(I_b + I_{in})$. Then, the gate terminal is disconnected from the drain and left open. As a result, its gate–source voltage is sampled and held by its parasitic gate–source capacitance C_{GS} at its potential at the time of opening $V_{GS}(I_b + I_{in})$. Hence the resulting output current I_{out} is equal to I_{in}.

Figure 4.13 shows a less idealized implementation of the copier. The main non-idealities in this more realistic current-copier circuit include attenuations due to finite in/out impedance ratios, charge injection coming from the MOS switches, and the discharge (which one hopes is slow) of the parasitic capacitor C_{GS} that leads to a droop in I_{out} over time. The main cause of this discharge is the inevitable presence, at the

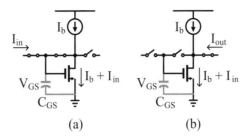

Figure 4.12 The principle of operation of a current copier: (a) during "track" mode V_{GS} follows the input; (b) during "hold" mode, the sampled input current is stored by means of the charge in C_{GS} and the corresponding V_{GS} forces an output current that should ideally be $I_{out} = I_{in}$ at the time of sampling.

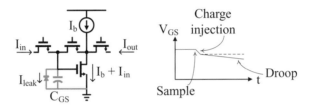

Figure 4.13 A more realistic current copier. When the input is sampled, the opening of the switches causes charge injection into the memory transistor's C_{GS}, hence altering the sample. Moreover, the reverse-biased drain-to-bulk and source-to-bulk junctions of the switches tied to the gate of the memory transistor can be modeled as a reverse-biased diode in parallel with C_{GS}. This diode introduces a charge leak that discharges C_{GS} and hence causes a droop of V_{GS} and I_{out}.

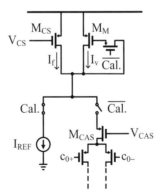

Figure 4.14 The principle of operation of the current-source calibration. During calibration a reference current $I_{REF} = I_u$ is imposed on M_{CS} and M_M in parallel. So any deviation of the fixed current I_f from its nominal value is absorbed by the variable current I_v and memorized by M_M. During normal operation the gate of M_M is open and I_u is routed to the current-steering switches.

memory transistor gate node, of the reverse-biased junctions introduced by the MOS switches used to sample V_{GS}.

The current-copier technique can be applied to the calibration of current sources as exemplified in Fig. 4.14. Here, during calibration, a reference current I_{REF} is superimposed on an MOS current source implemented as the combination of a (large) device M_{CS} with a fixed current I_f and a (small) device M_M with a variable current I_v. Let's assume that the reference current I_{REF} is set to the desired value I_u and that I_f is designed to have a nominal value slightly lower than I_u. Then, when I_{REF} is forced on the MOS current source, any negative/positive current deviation (e.g. arising from mismatches) on the sum $I_f + I_v$ will force the value of I_v to change so that $I_{REF} = I_u = I_f + I_v$ can hold (assuming, of course, that the range of I_v can accommodate it). Once the calibration phase is completed, the PMOS switch gating M_M opens and the total calibrated current $I_f + I_v$ is routed toward the current-steering switches.

If all the unity-current sources for a DAC array are implemented as just described, then a single reference current I_{REF} can be used to set their currents, one at a time. In this way, all currents will be calibrated relative to a unique reference and, to within the limits of this process, can be made equal to I_u.

If the calibrated currents could be held constant over time then the calibration could be done once (e.g. at start-up) and then the DAC could be operated after that (this would be a foreground calibration). Unfortunately, due to the above-mentioned discharge of C_{GS} a periodic "refresh" of the calibrated currents is necessary. So, while the DAC is operating with the calibrated currents, periodically, one at a time, each current source is disconnected from the DAC's output to be calibrated and an additional, previously calibrated current source (sometimes referred to as a "spare" current source) replaces the source under calibration so that the process is transparent to the DAC's operation. This refresh process continues periodically in the background. As explained in Section 2.5, one advantage of background calibration is that it will also allow the circuit to track I_{REF} over any changes, for example due to process, supply, and temperature variations, during the DAC's operation.

This principle can be used to directly calibrate all the nominally identical currents of a thermometric DAC; for example, that is typically the case for the MSB segment in a DAC such as the one depicted in Fig. 4.4. For the LSB segment, typically, its total current (nominally equal to I_u) will be also calibrated to be equal to I_{REF}, for instance by temporarily shorting all the cascode currents in the splitter into I_{REF} during the calibration phase. Insofar as the splitter's current ratios are concerned, intrinsic matching of the cascode transistors will need to be achieved. But, as explained earlier, the matching requirement is greatly relaxed in this structure.

4.2.4   Advanced calibration techniques

This calibration approach has recently been generalized in the "nested background calibration" [89] developed for a segmented DAC composed of an MSB, a ULSB, and an LLSB segment. In order to avoid major discontinuities occurring when bit changes cause transitions between segments, the nested calibration trims the segment boundaries. In the MSB–LSB segmented DAC, the sum of the currents of the LSB segment was calibrated to the nominal current I_u of one of the elements of the upper (MSB) segment. Similarly, in the MSB–ULSB–LLSB DAC with nested calibration algorithm, the sum of the currents in a lower segment is calibrated to the smallest element in the next higher (previously trimmed) segment. Again, the current splitters within each segment rely on intrinsic matching.

For example, as described in [89], in an $M-U-L = 6-2-5$ segmented DAC, first the $2^M = 64$ MSB I_u unit cells (namely, the entire MSB array plus one dummy MSB cell) are calibrated with respect to I_{REF}. Then the sum of $2^U = 4$ ULSB cells (which is equal to I_u) is calibrated with respect to I_{REF}. After that, the total current of the entire ULSB array ($2^U - 1 = 3$ ULSB cells) plus the sum of the current of the entire LLSB array (plus one dummy LSB current), which is again nominally equal to I_u, is calibrated with respect to I_{REF}. As can be seen in this example, dummies are used to complete the total required currents during calibration. Also, just as described before, if the calibration runs in the background while the DAC operates, previously calibrated spare cells/segments are employed to synthesize the output while identical sections of the current arrays are being calibrated.

As discussed above, all of the background calibration algorithms periodically "refresh" the currents in the array to re-set their value to the desired reference. Let us consider the circuit shown in Fig. 4.14. The act of disconnecting the cascode M_{CAS} from the output steering switches to connect it to I_{REF} through additional switches inevitably creates a disturbance visible at the output node. This happens periodically at the refresh rate f_{CAL} of the calibration process, hence resulting in undesired tones in the DAC output. One way to deal with that is to continuously vary the refresh rate in a random fashion within a prescribed range $f_{CAL1}-f_{CAL2}$. On doing this the undesired output tones are "scrambled" into a broadband noise-like output component, ideally disappearing within the noise floor [89].

Using the current-copier method described above, due to the charge leakage in C_{GS}, the calibration cycle needs to be repeated periodically to refresh the correction currents. An alternative to that is to store the correction in a longer-lasting fashion, avoiding or postponing the need for a refresh cycle. For example, in [298] the difference between the pre-calibrated current and the reference current is digitized using an auxiliary ADC, called cal-ADC; this datum is then stored in RAM and, using an ancillary DAC, called cal-DAC, the correction current is added to the calibrated current source. A diagram illustrating this principle is shown in Fig. 4.15.

Both the cal-ADC and the cal-DACs do not need to sample and convert rapidly since the required correction currents are essentially constant (in the case of a foreground calibration) or slowly varying (for background calibration). They also do not need to have high resolution/accuracy since the correction/error current is only a small fraction of the calibrated current to start with. Therefore, for example, a very-small-area and power-efficient successive-approximation ADC or an incremental ADC would be a very suitable choice for the cal-ADC. A single such ADC can be used to determine the digitized error-corrections for all of the calibrated currents. Each digitized error-correction data word will require only a few bits. These words require small RAMs and, likewise, simple, small, low-accuracy cal-DACs, one for each calibrated element. So a one-word RAM cell and one cal-DAC are lumped with each calibrated current

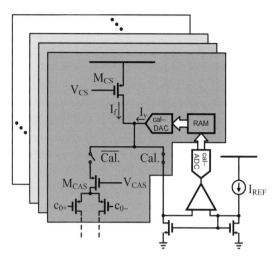

Figure 4.15 Current calibration using a mixed-signal calibration loop [298]. © 2003 IEEE.

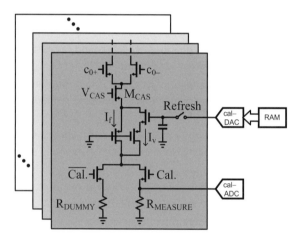

Figure 4.16 A floating current source [316]. © 2000 IEEE.

source. For example, the 18 b 400 MS/s DAC discussed in [298] uses a single 6 b SAR to determine the digital corrections, and 129 comparably low-resolution, very-low-area, cal-DACs for the 127 unary currents in the MSB segment, plus one for the combined intermediate–lower LSB segments and a last one for a biasing section.

   A somewhat similar approach has been proposed in [316], where a $\Sigma\Delta$ cal-ADC is used to measure the mismatch errors and a single cal-DAC is used to apply the correction (5%–10% of the total unary source current) to the calibrated sources. The principle of operation is depicted in Fig. 4.16. Here only one cal-DAC is used because, similarly to the current-copier approach, the correction currents for each cell are held locally with each calibrated cell using the combination of a capacitor and transconductor (see the branch below the cascode transistor M_{CAS} in Fig. 4.16). The measurement cycle and the correction cycle occur independently and simultaneously and without disconnecting (and replacing) the calibrated sources from the output. Each source is sequentially measured against a reference (which is set to be the sum of the intermediate/lower LSB segment current, so that gaps between the segments are canceled) by sensing the current from the supply side (using $R_{MEASURE}$ in Fig. 4.16) rather than the output side, hence avoiding a direct coupling with the output. The correction data is again stored in a RAM. At the same time the single cal-DAC periodically restores the charge in the capacitors, converting into a voltage the calibration data in the RAM. Because of the way the current source is built and sensed, this scheme has been named a "floating current source."

Finally, in [317], the idea of a background calibration using floating current sources is further explored in a low-voltage context and, once again, with a fully analog calibration loop (not requiring digitization). The analog calibration loop eliminates the digitization overhead while the use of floating sources has the advantage of minimizing the coupling to the output.

So far we have considered schemes in which the current to be calibrated is sensed at the drain of the current-source devices M_{CS} and M_M. For example, that is the case for the scheme of Fig. 4.14. However, since to set the current of M_M during calibration this device is diode connected, this approach has two fundamental limitations. First of all, it limits the voltage headroom since it does not allow biasing of the cascode M_{CAS} with a source potential closer to the supply (yet still keeping M_M and M_{CS} in saturation). Second, depending on how the cascode is biased, the drain potential for M_M–M_{CS} might vary on going from calibration (when M_M is diode connected) to normal operation (when it is imposed by M_{CAS}'s source potential), hence introducing a systematic variation between the calibrated value of I_u and the sourced current as a result of channel-length modulation. Both issues are addressed if the cascode M_{CAS} is included in the calibration. That is addressed, for example, in [318, 319], where an analog calibration loop is introduced around the entire M_M–M_{CS}–M_{CAS} structure as shown in Fig. 4.17.

In Fig. 4.17, the trapezoid marked with R on the right-hand side represents a transresistance amplifier, which serves two main purposes. It closes the feedback loop around the cascoded current source, providing a low-impedance point where the reference calibration current I_{REF} is injected. It also implements a voltage level shift between the drain of M_{CAS} and the gate of M_M, keeping the calibrated devices at their desired biasing point. In fact, designing the transresistance amplifier so that its input potential tracks the drain voltage of M_{CAS} during normal operation [319] eliminates any residual changes in I_u between calibration and operation mode (although introducing M_{CAS} alone into the loop has already diminished this current error by a factor approximately equal to M_{CAS}'s intrinsic gain compared with the previous scheme). As shown in Fig. 4.17, only one such transresistance amplifier is required in order to sequentially calibrate all current cells. A similar approach was subsequently presented in [89] as well.

All the previous calibration techniques correct for static mismatch in the current sources only. Therefore these can be used to improve the DNL and INL but have only

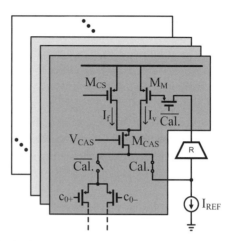

Figure 4.17 Calibration including the cascode transistor.

indirect benefit regarding the dynamic performance of DACs (a topic discussed in subsequent sections). The indirect benefit consists of the fact that, if calibration is used to correct for DC current mismatch originating from the mismatch in the MOS devices setting such currents, then the designer has far greater freedom to size all relevant devices for high-frequency performance. Namely, the corresponding sizes and quiescent points can be set to lead to the shortest possible time constants along the signal path (as explained more precisely in Section 4.3).

Recently, however, more sophisticated calibration techniques accounting also for some of the dynamic performance limitations have been proposed. For example, the calibration technique proposed in [320] is based on the following considerations. Let us consider the unity current $I_{u,j}$ provided by one (e.g. the jth cell) of the unary sources in the MSB segment of a DAC and the reference current I_{REF}. The static mismatch $E_j = I_{u,j} - I_{REF}$ is what has been nulled using the previously described calibration techniques. However, when, during normal operation, the jth current is switched between the positive and negative outputs of the DAC, due to transients and timing-skew mismatches between this unary cell and the reference cell, timing errors add up to amplitude (DC) errors E_j. In principle, one way to assess the dynamic error could then be to alternately switch $I_{u,j}$ and I_{REF} between the positive and negative outputs at a frequency f_m and to measure their *dynamic* difference $e_{f_m,j} = i_{u,j} - i_{ref}$ under this condition.[8] If we take the FFT of the dynamic mismatch waveform $e_{f_m,j}$, this periodic signal will have a fundamental component and harmonics. Each of these tones can be represented by a phasor with its own I and Q components. So, for the first harmonic, at frequency f_m, the mismatch phasor can be represented by $E_{f_m,j} = (I_{f_m,j}, Q_{f_m,j})$; the second-harmonic phasor will be $E_{2f_m,j} = (I_{2f_m,j}, Q_{2f_m,j})$ etc. The first and second harmonics bear the largest power and so the higher-order harmonics are neglected. Since these two tones are used to represent the difference between each individual source and the reference, they can be

---

[8] Note the use of lower-case letters to emphasize the time-varying component of these currents.

considered as representations of a "dynamic DNL." In line with that, a "dynamic INL" is defined as

$$\text{inl} = \sqrt{\left|\sum_{j=0}^{2^{M-1}} E_{f_{\mathrm{m}},j}\right|^2 + \left|\sum_{j=0}^{2^{M-1}} E_{2f_{\mathrm{m}},j}\right|^2} \qquad (4.13)$$

The algorithm proposed in [320] estimates the powers of $E_{f_{\mathrm{m}},j}$ and $E_{2f_{\mathrm{m}},j}$ for all unary cells; then it automatically changes the switching sequence of the cells with the objective of minimizing the above-defined "dynamic INL." Hence, this approach can be thought of as a dynamic extension of the static techniques based on switching sequences discussed in Section 4.2.2. Although, in addition to accounting for dynamic errors (while in the methods of Section 4.2.2 only static errors are explicitly targeted), with this approach (a) dynamic errors are actually measured and (b) an optimal switching sequence is derived accordingly and applied to change the DAC's operation.

It is also important to note that once the optimal switching sequence has been found and applied to the DAC no scrambling of the cells occurs. Distortion is hence systematically minimized. That is in contrast with DEM techniques (discussed in Section 4.5) insofar as here distortion is not scrambled into a broadband set of spurious tones that degrade the noise spectral density.

A couple of additional important remarks are in order. First of all, the dynamic error phasors are determined at a single frequency f_{m}; since these phasors will vary for different choices of f_{m}, the optimization performed is in general valid for f_{m} only and, therefore, different results, in terms of DAC dynamic performance, have to be expected depending on the choice of f_{m}. It is intuitive that for low f_{m} the calibration performs similarly to a more traditional static calibration. Second, regardless of the fact that this is implemented as a foreground or background calibration (using a pre-calibrated spare cell in operation to replace the one being calibrated), the comparison between the jth cell and the reference is done in isolation from the output. Hence dynamic issues arising from the effect of the output to the jth cell itself are not accounted for (this will be discussed further in Section 4.3). Lastly, although here mismatches are actually estimated, on the other hand, just like for the optimized static switching sequences described in Section 4.2.2, there is no actual correction of the sources of mismatch (including the DC current values), but rather a minimization of their global effect (the dynamic INL of Eq. (4.13)) through an optimal sorting of the switching sequence.

An actual circuit implementation of this approach can be found in [320], where (a) additional steering switches are cascoded to the current cells to isolate them for the mismatch estimation, (b) a chopping circuit followed by a continuous-time $\Delta\Sigma$ ADC is used, as a zero-IF scheme, to digitally estimate the mismatches in a sequential order, and (c) the sorting algorithm determining the optimal switching sequence is implemented off-chip and the resulting sequence is then applied to the programmable thermometer decoder for the MSB segment of the DAC.

Another dynamic calibration approach that bears some key commonalities with the latter method is proposed in [321, 322]; this time, however, the calibration is performed in the background without disconnecting the cells from the output or using spare cells; also

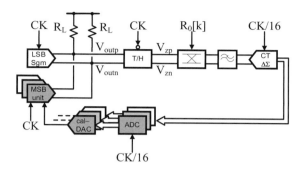

Figure 4.18 A calibration loop for compensation of dynamic effects [321]. © 2011 IEEE.

the switching sequence is not changed, while the current sources are actually controlled in order to compensate for the effect of the mismatches.

A high-level scheme illustrating the idea is shown in Fig. 4.18. The DAC is composed of a binary weighted LSB segment and a unary MSB segment, depicted on the left-hand side of Fig. 4.18, feeding the output current to load resistors R_L. Each unary source in the MSB segment is calibrated to match the total current of the LSB segment (removing segment boundary discontinuities and, also, implying that the total current of the LSB segment is the reference current). The DAC is a return-to-zero (RTZ) DAC with a 50% duty cycle. The calibration runs in the background and is active in the half-period during which the DAC output is being zeroed. The output zeroing is accomplished by forcing half of the current sources (half of the MSB unary sources) to steer their current to the positive output while the other half (in fact, the remaining MSB unary sources plus the total LSB segment) is steered to the negative output. The order in which these are selected is determined by a pseudo-random-number generator (PRNG) implemented using a linear feedback shift register [322]. The turning-off transient (and hence also any non-zero residue due to the current mismatch) during the output-zeroing phase is sampled using a track and hold (T/H) and the contribution due to the MSB unary element to be calibrated is extracted from the output by using the chopper, driven using the pseudo-random binary correlation sequence $R_0[k]$ corresponding to that specific unit element for the zeroing phase of the output. That is digitized using a continuous time $\Delta\Sigma$ ADC, the output of which is used to update an accumulator corresponding to the calibrated unary element, which ultimately drives the calibration DAC for that element. As usual, all unary elements are sequentially calibrated and the loop is repeated, providing a periodic refresh of the MSB currents.

Although the calibration acts to change the DC currents of the MSB unary sources and, therefore, there is no explicit corrective action on dynamic errors such as timing-skew delays between the unary elements, this approach, in addition to measuring the residual non-zero output (resulting from uncalibrated DC current mismatches) also measures the transient response (during zeroing) of the output, with all current-steering cells actively sourcing the load, and acts to equalize all contributions under these conditions. That is in stark contrast with all other methods previously considered, in which each element to be calibrated is instead isolated from the remainder of the array and only the DC current is measured. The results reported in [321] show a significant improvement in linearity not only at low frequency, but, in fact, sustained all the way to the Nyquist frequency and becoming limited, as expected, by the timing-skew errors.

To conclude this section we will briefly mention that a possible alternative to calibration is trimming. Some recent examples of trimmed DACs have been reported in [323, 324]. The advantages and disadvantages of trimming versus calibration have already been discussed in Section 2.5.

4.3   Dynamic linearity

The previous sections discussed some of the imperfections leading to static linearity problems. These included

- DC current mismatch among the sources in the DAC array;
- finite output resistance of the current sources;
- ohmic ("$r \cdot i$") drops on power lines.

All these issues persist, extend and, in fact, can degrade further when the dynamic performance of DACs is considered. The previous analysis on the effect of output resistance of the current sources and the ohmic drops is extended by considering the corresponding impedances. The output impedances of the current sources, due to the effect of stray capacitances at various nodes constituting the current sources, decrease in magnitude for increasing frequency. Therefore the previously discussed code-dependent errors introduced by such finite resistances degrade further when the impedances become even smaller for rapidly varying signals.

The ohmic series drops on power supply lines also degrade as inductive components along these lines add up to the total series line impedance when such lines are traversed by dynamic signals. That is especially worrisome when the dynamic signal is correlated to the DAC output itself.[9]

In addition to all of the above, many new additional sources of linearity performance degradation become increasingly important as the frequency of the signals involved increases. These additional sources of problems include

- load imbalances between otherwise matched and/or weighted elements;
- switch and switch-driver mismatches;
- issues associated with the way each current is steered from the positive to the negative output;
- timing skews between the switching times of steering switches of different current sources;
- clock jitter (depends only on f_{out});
- coupling between otherwise disconnected circuit blocks (e.g. digital noise, supply and substrate noise, coupling between adjacent unary/binary elements); this is not negligible for high f_s even when f_{out} is low.

---

[9] Attention should be paid to the different influence of f_{out} and f_s. For example, even if f_{out} is relatively low, when f_s is large, high-frequency components are present in the current flowing along many lines and that causes the associated stray inductances to develop undesired local voltage transients.

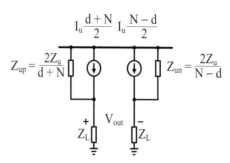

Figure 4.19 A linearized and simplified thermometer decoded DAC with finite source impedances [90]. © 2009 IEEE.

It should also be said that, although current-steering DACs have been widely designed and used for a few decades, meaningful published results and techniques to cope with these dynamic imperfections are somewhat recent and still relatively open to further analysis and full experimental verification. As result of all that, within the span of roughly 10 years (and only in limited part attributable to the progressing process nodes) Nyquist-rate CMOS DACs have been reported with distortion performance of SFDR $= 73\,\text{dBc}$ at around $10\,\text{MHz}$ in 1998 [306] (in $0.35\,\mu\text{m}$), at around $45\,\text{MHz}$ in 2000 [316] (in $0.35\,\mu\text{m}$), at around $200\,\text{MHz}$ in 2004 [311] (in $0.18\,\mu\text{m}$), and at around $500\,\text{MHz}$ in 2011 [321] (in 90 nm). Not only that, but also the output rate of 12–14-bit DACs has gone from the low hundreds of MS/s in the late nineties [325, 316], to breaking the GS/s barrier in the early–mid 2000s [311], and is continuing to inch higher and higher with sustained dynamic performance [90].

4.3.1   Finite output impedance

The issue associated with the finite output impedance has already been introduced in Section 4.2.1 [52, 90, 308, 326, 327, 328]. For the discussed example of a fully thermometric DAC it is possible to derive an expression for the contribution to distortion of this specific limitation in the case of a perfectly balanced circuit and differential output [90]. Let us for convenience represent the DAC's digital input D_{in} with the signed input variable d, such that $-N \le d \le N$ (where $N = 2^n$ is the number of current sources). The small-signal linearized circuit shown in Fig. 4.8 can be reduced to the simplified circuit shown in Fig. 4.19.

As previously discussed, the ideal converter current goes mostly to the load Z_{L} as intended but also partly to the code-dependent impedances Z_{up} and Z_{un}, resulting in distortion. The output voltage phasor is therefore [90]

$$V_{\text{out}} = Z_{\text{L}} N I_{\text{u}} \left[\frac{d}{N} + \left(\frac{d}{N} \right)^3 \left[\frac{Z_{\text{L}} N}{2 Z_{\text{u}}} \right]^2 + \cdots \right] \tag{4.14}$$

In this equation there is clearly visible a first term that is linearly dependent on the input code d representing the intended conversion output signal. There is, however, also a

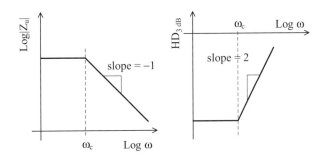

Figure 4.20 The frequency dependence of the magnitude of the unity source impedance and of the corresponding HD_3. The corner frequency is $\omega_c = 1/(R_u C_u)$. Adapted from [90].

power series with the odd-harmonics contribution to the output. The first (cubic) term of that is the third-order distortion and hence the third-harmonic distortion for a full-swing output is promptly derived:

$$HD_3 = \left[\frac{|Z_L|N}{4|Z_u|} \right]^2 \tag{4.15}$$

In general, the higher odd-harmonic distortions due to this phenomenon are given by [326, 328]

$$HD_k = \left[\frac{|Z_L|N}{4|Z_u|} \right]^{k-1} \tag{4.16}$$

In the common case in which the load is primarily resistive ($Z_L = R_L$, commonly that is either 50 Ω or 70 Ω) and the unity current source impendance can be represented as having a dominant pole ($Z_u = 1/(1/R_u + j\omega C_u)$) [329], HD_3 will degrade quadratically with frequency as shown in Fig. 4.20 [90]. In general, HD_k will degrade with a frequency slope equal to $k - 1$ [86].

The severity of this type of performance degradation can be realized with a numerical example. Let us consider a 14-bit DAC segmented as 7–3–4. The thermometer decoded segment has then $2^7 - 1 = 127$ unary cells, plus one more unary current I_u is current-split to implement the upper and lower LSB segments. Hence we can use the previously discussed thermometric model for $N = 128$. If $R_L = 50 \Omega$ and the desired maximum distortion is $HD_3 = 0.000\,316$ (namely -70 dB), the minimum unary impedance is found using Eq. (4.15) to be

$$|Z_u| \geq \frac{R_L N}{4\sqrt{HD_3}} = \frac{50 \cdot 128}{4 \cdot \sqrt{0.000\,316}} \simeq 90\,k\Omega \tag{4.17}$$

Solving the corner frequency $\omega_c = 1/(R_u C_u)$ relation for a desired maximum output frequency of $f_{out} = 500$ MHz leads to a maximum tolerable parasitic capacitance $C_u = 1/(2\pi f_{out} R_u) = 1/(2\pi \cdot 500\,\text{MHz} \cdot 90\,k\Omega) \simeq 3.5$ fF. The latter is unrealistically low and gives the reader a sense of the importance of this problem.

As was seen during the derivation of Eq. (4.15), the issue does not directly stem from the presence of the parasitic capacitances C_u per se. As can be seen in Fig. 4.19,

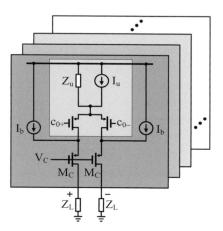

Figure 4.21 A biased cascode scheme to minimize the distortion due to finite switching impedance [90]. Adapted from [90].

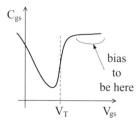

Figure 4.22 The gate–source capacitance of an MOS transistor as a function of the bias point.

the distortion originates from the fact that such impedances/capacitances dynamically change which output node they end up connected to, as a function of the switches' state, hence dynamically "stealing" currents to the output in a code-dependent fashion. As always, the key word is "code-dependent," as it translates into distortion.

A solution to this problem, recently proposed in [90], consists of making the capacitance contributed to each of the output terminals independent (at least to first order) from the status of the switches. That is accomplished by augmenting the traditional two-switch cell (highlighted by a lighter shaded rectangle in the depicted scheme) with a biased cascode M_C as shown in Fig. 4.21.

If the biasing current I_b is too low (or zero [330]) then, when the corresponding switch is off, the cascode too will turn off and the corresponding capacitance contributed at its drain will be dramatically different from that contributed when it is on and carrying the current $I_u + I_b$. So, taking advantage of the nonlinear characteristics of the gate–source capacitance of MOS transistors [5, 33], shown in Fig. 4.22, it is sufficient to provide a current I_b that will keep M_C's C_{gs} within the relatively constant region on the right-hand side of this plot even when the corresponding series switch is off. A bias current of the order of 1%–2% of I_u has been shown to be sufficient for this purpose

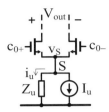

Figure 4.23 A simplified schematic representation of one of the DAC current-steering cells.

[90]. In this way, the capacitance seen at the drains of the cascodes M_C is approximately constant regardless of the status of the switches and hence leads to a finite but fairly code-independent output impedance, dramatically reducing the associated distortion contribution.

4.3.2 The common source node of the steering switches

One of the most delicate nodes for the analog part of a DAC is the common source node of the current-steering switches, namely node S in Fig. 4.23 [327, 331]. We have already seen in the previous section that if this node's capacitance, after "reflection" from the switches, shows up alternatively in the negative or positive output in a code-dependent fashion this can lead to high-frequency harmonic distortion. There are, however, more important reasons to keep this node as isolated from the output as possible and fairly quiet in terms of signal swing.

Owing to the presence of Z_u at S, any voltage swing v_S at this node leads to current i_u passing into Z_u. This current i_u adds to the desired current I_u at S and hence ends at the DAC output after passing through the switches. If i_u is signal/code-dependent then it introduces output distortion. To be clear, Z_u in Fig. 4.23 is the lumped contribution of the actual current source circuit implementation used to generate I_u but it also includes the contribution of the two switches and their associated parasitics (e.g. primarily their C_{gs}s).

If the cell is isolated (i.e. no meaningful coupling from neighboring circuits occurs) then the voltage swing v_S can result from two sources: (a) the output swing V_{out} itself and (b) the control signals c_{0+} and c_{0-} driving the switches. Both of these sources couple into S through the switches, so the design must focus on dealing with (a) minimizing the corresponding coupling mechanism and/or (b) the nature of the involved signals themselves.

First of all, let us consider the coupling mechanisms through the switches. The coupling of $V_{out} = V_{outp} - V_{outn}$ into S occurs primarily because of finite isolation between the drain and source of the switch that is turned on.[10] At low frequency and for small

---

[10] There is also some capacitive coupling through the C_{ds} of the off switch; however, at least at moderate frequency, that is definitely negligible compared with the much more direct coupling through the turned-on switch.

signals the single-switch coupling consists of an attenuation amounting to the intrinsic gain $A_{sw} = g_m/g_{ds}$ of the on switch (e.g. the one steering the current to the positive output terminal at voltage V_{outp}) if the latter is kept in saturation:

$$v_S \simeq V_{outp}/A_{sw} \qquad (4.18)$$

So, for example, if $A_{sw} = g_m/g_{ds} \sim 15$ then, by observing the node S of various current cells, while the DAC is synthesizing V_{out}, it will be possible to see a 15-fold attenuated copy of the positive output node voltage V_{outp} for the cells with $c_{0+} > c_{0-}$ and a 15-fold attenuated copy of the other node voltage V_{outn} for the other cells.

This attenuation will dramatically drop if the switch is driven into triode (instead of saturation) either by design or as a result of a large output swing. In this case the on switch can be modeled as a resistor (the channel resistance) and a large v_S causes a large dynamic distortion current i_u. Because of that, from now on we will assume that the DAC has been designed to keep the on switches in saturation under all circumstances.

The switches' gain A_{sw}, just like any gain, will degrade at high frequency. The sizing of the switch should be optimized for the intended frequency of operation, taking into consideration both the parasitics degrading its high-frequency performance and the fact that, although a minimum length $L = L_{min}$ maximizes g_m, it also leads to relatively small g_{ds}^{-1} due to channel-length modulation and short-channel effects (in other words, $L = L_{min}$ isn't likely to be an optimal choice) [33].

Considering a higher-frequency case, as one switch turns on and the other one turns off, v_S swings from the attenuated copy of the node voltage of one of the two output terminals to the attenuated copy of the other node voltage. Therefore the total capacitance $C_u = 1/(\omega \cdot \text{Im}(Z_u))$ at S varies its charge accordingly through i_u. It can be proved [86, 327] that, for an N-cell DAC and a sine-wave signal at frequency f, the corresponding second- and third-harmonic distortions are

$$HD_2 = \frac{\pi f Z_L C_u N}{2 A_{sw}} \qquad (4.19)$$

$$HD_3 = \frac{\pi f Z_L C_u N}{4 A_{sw}} \qquad (4.20)$$

On combining Eqs. (4.15) and (4.20) it is possible to see that for frequencies lower than

$$f_3 = \frac{A_{sw}}{\pi Z_L C_u N} \qquad (4.21)$$

the source node S capacitance switching effect of Eq. (4.20) dominates the third-order distortion, while for higher frequencies the output-impedance-induced distortion described in Section 4.3.1 dominates.

Using a "telescopic design"[11] for the cascoded (and/or regulated cascoded) current sources to implement the source I_u allows a minimal contribution to C_u [298]; however, again, one needs to also bear in mind the contribution to C_u of the gate–source capacitance C_{gs} of the switches. Also, note that, although a dominant-pole model for Z_u has

---

[11] Namely one with smaller and smaller drain capacitances as the stacked transistors approach the output node.

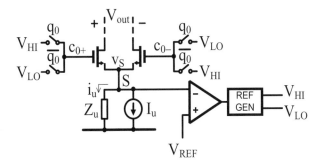

Figure 4.24 A conceptual diagram for boosting A_{sw} by means of negative feedback and control signal $c_{0+/-}$ modulation.

been assumed so far, in general, depending on the implementation of the current source, there might be a plurality of poles with none quite dominating for all frequencies [326]. Although the maths describing the distortion will change accordingly, the underlying principles don't change too much [86].

Moreover, the technique using the biased cascoded switches [90] described in Section 4.3.1 addresses both distortion mechanisms since it magnifies the attenuation of $V_{outp/n}$ at v_S by means of the cascode M_S in series with the switch in addition to also mitigating the code-dependent switched-impedance problem as discussed previously.

Clearly, in all these cases, any additional level of cascoding needs to be carefully considered in the light of voltage headroom issues in nanometer CMOS processes.

Finally, the switch's intrinsic gain A_{sw} can be boosted for moderate signal frequencies by means of local negative feedback sensing the node S and modulating the turn-on switch's control signal c_0 to counter the swing of v_S as described in [332] and shown in the conceptual scheme of Fig. 4.24. In this technique the absolute difference $\delta_V = V_{HI} - V_{LO}$ between the switch turn-on voltage V_{HI} and the switch turn-off voltage V_{LO} is kept constant; however, these two voltages are locally regulated through the feedback amplifier so that v_S is driven to stay as close as possible to V_{REF}.[12] The approach works well for output frequencies within the bandwidth of the regulating loop and does not limit the voltage headroom because no additional transistors are stacked up [332].

4.3.3   Switches and their drivers

The last technique described in Section 4.3.2 represents a first example of a switch driver. As we stated in the same section, another possible source of code/signal-dependent disturbance to v_S (and hence distortion) is represented by the coupling of $c_{0+/-}$ to node S. In the case of the previously discussed coupling of $V_{outp/n}$ into v_S nothing really can be done on $V_{outp/n}$, so design acts to mitigate the coupling. On the other hand, in the case

---

[12]   In reality, it is sufficient to modulate V_{HI} alone to control v_S as long as V_{LO} is kept sufficiently low to insure that the corresponding switch is off as desired.

of the coupling of $c_{0+/-}$ into S, the designer has control over the nature of $c_{0+/-}$ and can therefore implement proper countermeasures as explained below.

A number of considerations can be made on the switches' control signals $c_{0+/-}$. The current-steering cell of Fig. 4.23 is nothing more than a classic differential pair, with the steering switches as the "input pair." As such, for it to work optimally, the control signals $c_{0+/-}$ need to be properly balanced, and must have a constant common mode level $c_{0CM} = (c_{0+} + c_{0-})/2$ setting the quiescent point of the node S to $V_S = c_{0CM} - V_{GSsw}$, where V_{GSsw} is the switch M_{SW}'s gate–source voltage corresponding to the full current I_u:

$$V_{GSsw} = V_T + \sqrt{\frac{2I_u}{\beta}} \tag{4.22}$$

If c_{0CM} is indeed constant both when $c_{0+/-}$ are going through the on–off transition and when they are going through the off–on transition (while I_u is gradually steered between one output node and the other) as well as when they have fully settled to the desired state (and hence I_u is fully steered) then, in principle, node S sees no coupling from $c_{0+/-}$, namely $v_S = V_S$ (ignoring $V_{outp/n}$ coupling) and no distortion is introduced as a result. In practice, even if c_{0CM} is truly constant at all times, a very small v_S variation (and a glitch) will show up at S as a result of the MOS square law relating the drain current of the current switches to their gate–source voltages,[13] as a result of stray capacitances, device mismatch, and second-order effects.

As for the amplitude δ_V of the differential control signal $c_0 = c_{0+} - c_{0-}$, this requires only that most (i.e. depending on the required resolution) of the current I_u is steered onto one output branch when the transition has been completed. Since the differential transfer characteristic $i_d = i_{d+} - i_{d-} = G(c_0)$ of the switch pair is the large-signal function [5]

$$i_d = I_u \sqrt{\frac{\beta c_0^2}{I_u} - \frac{\beta^2 c_0^4}{4I_u^2}} \quad \text{for} \quad |c_0| \le \sqrt{\frac{2I_u}{\beta}} \tag{4.23}$$

then, on setting $\delta_V = \sqrt{2I_u/\beta}$, in theory for $c_0 = \pm\delta_V$ all of the current I_u has been fully steered. In practice, that is a good initial guess but the minimum δ_V can be found by simulation, also due to the fact that nanometer-scale MOS transistors don't strictly follow the square law.

Given all that, it is possible to design suitable differential drivers for the switches such as the circuitry exemplified in Fig. 4.25. Here, with suitable sizing of the tail source I_{dr} and the resistors R_1 and R_2, it is possible to convert the digital rail-to-rail signal q_0 into a proper differential control signal $c_{0+/-}$ with desired common mode level and

---

[13] In practice this effect is tolerable (and will be neglected for simplicity in the remainder of this discussion) since the current switches are typically of small size and therefore, during this transition, they rapidly move to conducting high current density, a region of the parabolic law where the slope doesn't vary too much and the law is almost linearized. This leads to a rather reduced swing at S. Furthermore, in finer nanometer processes, due to strong velocity saturation, the power of 2 in the MOS square law is more realistically approximated by a fractional number between 1 (linearity) and 2 (the square law). In other words, the swing at S due to this law reduces considerably. In theory it is zero if the exponent is 1.

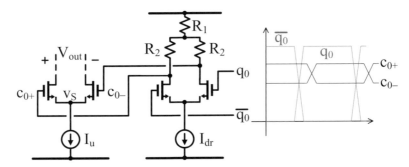

Figure 4.25 The differential driver scheme.

amplitude.[14] In fact, multiple CML stages might need to be interposed between the rail-to-rail logic and the CML drivers to improve the balance of the signals. Moreover, care should be taken in the design of such drivers to insure that the high-to-low and low-to-high transitions of $c_{0+/-}$ are even.

A possible downside of such a type of "CML" driver is the relatively high static power consumption (due to the fact that I_{dr} is flowing at all times) and area (primarily due to the resistors) [334, 335]. However, it should be said that, for relatively high speed, CML circuits are known to be as power efficient as (or even better than) more traditional static CMOS logic [336].

In order to mitigate both the power consumption and area issues in the differential drivers, other approaches have been pursued. Similarly to the approach shown in Fig. 4.24, two reference levels V_{HI} and V_{LO} can be created so that the steering switches can be turned on and off by forcing their gate terminals to these levels as needed. Two options are possible: a single ("global") reference generator for V_{HI} and V_{LO} complemented by local buffers supplying these voltages to the local switches; or multiple "local" generators associated with each pair (or a "local" group) of switches. Expectable trade-offs among power consumption, pair-to-pair coupling, layout routing, driving capability, and area need to be considered for the architecture and segmentation chosen [63, 298].

An example of a reference generator suitable for a P-type DAC is depicted in Fig. 4.26 [63]. In this scheme it is assumed that a low-impedance-output common mode reference $V_{outCM} = (V_{outp} + V_{outn})/2$ is available and sets V_{LO} (which, for a P-type DAC, is the level that turns on the desired switches). The circuit of Fig. 4.26 develops the desired voltage level shift δ_V needed to turn off one of the switches (by setting its gate at $V_{HI} = \delta_V + V_{LO}$) when the other one is fully on (i.e. with its gate at $V_{LO} = V_{outCM}$).

In principle the driving scheme for the steering switches could be the one shown in Fig. 4.27 using $M_{N1}-M_{N4}$ to connect the steering switches to the reference generator. However, when the steering switches' gates become tied to V_{HI} and the charges at those nodes need to change so that the new voltage level is established, then, due to the finite

---

[14] In [333] the input range of the large-signal characteristic of the switching pair Eq. (4.23) is extended by degenerating the pair in order to use a driver with larger swing instead.

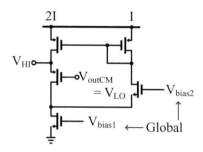

Figure 4.26 A V_{HI}/V_{LO} reference generator [63] for a P-type DAC. This generator, consisting essentially of a P-type source follower, is intended for an individual switch pair and creates V_{HI} from a supplied V_{LO}. The biasing voltages V_{bias1} and V_{bias2} can be global [63]. © 2007 IEEE.

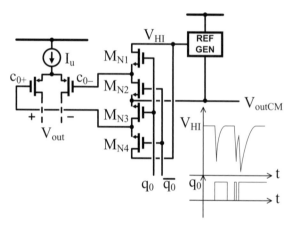

Figure 4.27 A simple driver using the reference generator of Fig. 4.26. When $q_0 = 1$ then $c_{0-} = V_{HI}$ through M_{N1} and $c_{0+} = V_{LO}$ through M_{N3}. When $q_0 = 0$ then M_{N2} and M_{N4} set $c_{0+} = V_{HI}$ and $c_{0-} = V_{LO}$ [63]. © 2007 IEEE.

driving impedance of the reference generator, a switch gate voltage $c_{0+/-}$ transient develops as shown in the small insert plot of Fig. 4.27. If the switching activity is high enough then V_{HI} and the switch gate voltage $c_{0+/-}$ might not be settled yet when a new switch-state change happens. Therefore the reference generator is unable to re-establish the desired V_{HI} ahead of the new switching transition and a code-dependent memory effect (effectively an inter-symbol interference problem) on the generated voltage leads to a new introduction of dynamic nonlinearity at the DAC output.

This problem is addressed either by designing the reference generator with lower output impedance (in order to make the transient time constant shorter) or by selectively recharging the disturbed output reference node so that it can quickly recover. In the former case, power–area issues similar to those in the CML driver case can arise. An example of the second case is shown in Fig. 4.28 [63, 298]. In Fig. 4.28, a complementary

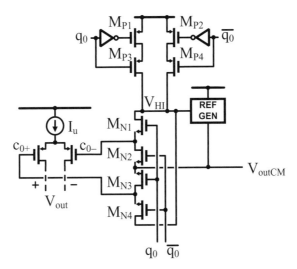

Figure 4.28 An enhanced version of the driver in Fig. 4.27. The extended structure with M_{P1}–M_{P4} and the two inverters transfers charge packets from the top supply to restore the charge lost by the reference generator when driving the steering switches, hence compensating for the code-dependent glitches in V_{HI} [63]. © 2007 IEEE.

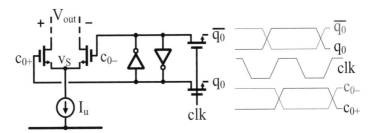

Figure 4.29 Simple latch driver scheme.

set of switches M_{P1}–M_{P4} provides the output reference node with the charge previously lost to drive the steering switches by selectively draining it from the positive supply.

An alternative to bounded-swing drivers, and, to some extent, a more traditional approach, is to directly drive the switches with a rail-to-rail fast latch such as the one depicted in Fig. 4.29.

In this case the state transition of $c_{0+/-}$ is intended to be the fastest possible. The design and layout of the switches and the latch need to be carefully done together, hence the term "swatch" (*switch–latch*) used by some [86].

When the control signals $c_{0+/-}$ swing rail-to-rail, as opposed to the previously discussed cases with bounded differential swing, there are several potential issues that need to be considered carefully. First of all, once the state transition is completed, the on switch's gate terminal is held at the positive supply voltage of the latch.

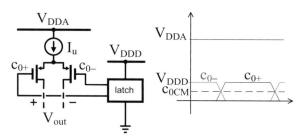

Figure 4.30 A P-type DAC with a lower supply latch.

Therefore, depending on the amplitude of the output voltage $V_{\text{outp/n}}$, the on switch may be (possibly temporarily) driven into triode, hence causing distortion (e.g. when $V_{\text{outp/n}} < V_{\text{Gsw}} - V_T$). Second, for both switches, a large and rapid swing $c_{0+/-}$ at the switches' gates results in a large charge injection (from the switch's channel) and feedthrough [330] (from the switch's C_{gd}) to the output node (and to the node S through the switch's C_{gs}), also introducing distortion. Finally, while the signals $c_{0+/-}$ are changing state, the corresponding common mode $c_{0\text{CM}}$ may vary from its desired value (discussed above), and it might be different from its final value at transition completed. Such variation of the common mode $c_{0\text{CM}}$ directly results in a variation of v_S, since, as explained above, v_S should always stay at $V_S = c_{0\text{CM}} - V_{\text{GSsw}}(I_u)$ in order to have $i_u = 0$. In fact, besides the obvious distortion introduced by varying the small-signal current i_u, a variation in v_S that pushes the top-most transistor of the current source from its intended saturation state into triode will cause a very large and sudden drop in I_u. This current will only begin to recover its original value after the on–off transition has been completed, returning to flow through the on switch. Again, this is a very large variation visible at the output, not a small-signal variation.

Using a "digital language" some authors describe this condition by stating that both switches turn off at the same time and so the current I_u cannot flow, when, in fact, as just described, the problem is that the switches are "yanked down" until they drive the current source out of its intended bias point, possibly turning it off.

There are basically two ways to address this problem: using drivers with reduced supply and, hence, once again reduced swing, or changing the crossing point of the $c_{0+/-}$ signals.

If a P-type DAC architecture is chosen then it is fairly straightforward to use a lower supply for the driver than for the current cell as shown in Fig. 4.30. That is particularly natural if a process with dual supply and corresponding lower-voltage/shorter-channel/higher-speed devices is used.

Conversely, in an N-type DAC this approach would result in either (a) lifting the bottom supply of the latch from ground to a higher voltage level or (b) lowering the bottom supply of the current cell to a negative supply.

The second way to address the problem is to change the crossing point of c_{0+} and c_{0-}. In this way, when their state transition occurs (and either switch is conducting at least some current), the corresponding $c_{0\text{CM}}$ is not very far from its intended level.

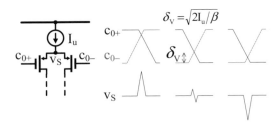

Figure 4.31 Variation of $c_{0+/-}$'s crossing point and its impact on node S.

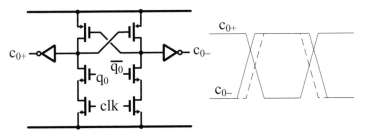

Figure 4.32 An example of a high-crossing-point driver.

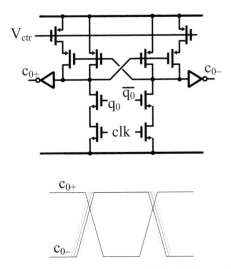

Figure 4.33 A variation on the previous driver circuit with the ability to adjust the crossing point.

Some examples clarifying the concept are depicted in Fig. 4.31 and an example of a switch driver with a skewed crossing point is shown in Fig. 4.32 [312, 328, 329, 330]. A variation of this circuit allowing one to change the crossing point by means of a control voltage is shown in Fig. 4.33.

Since the rise and fall time of these waveforms is meant to be as fast as possible, small timing skews can vary the crossing point (and its corresponding distortion performance), making it quite sensitive to PVT variations. A solution to automatically control it has

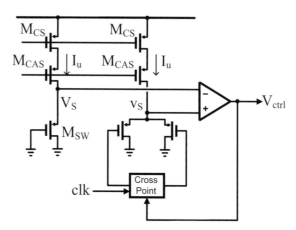

Figure 4.34 The concept for automatic control of the crossing point [311]. © 2004 IEEE.

been proposed in [311, 337]. It consists of monitoring the switching activity of the node S of an isolated replica of the current cell, automatically varying its crossing point to minimize the glitch at S, and using the resulting set point (V_{ctrl}) for all the drivers used in the actual DAC. The concept is illustrated in Fig. 4.34. On the left-hand side a partial replica of the current cell recreates the desired source voltage V_S. On the right-hand side a more complete replica has its switches turned on and off at the clock rate and its common source node voltage v_S experiences glitches depending on the crossing point. A regulating loop senses the switching common source node (v_S), compares it with the static node (V_S), and then adjusts the crossing point to minimize their difference. The resulting V_{ctrl} is then applied to all the controllable crossing-point switch drivers in the DAC array.

So far we have described different techniques that generate the control signals $c_{0+/-}$ aiming at minimizing the corresponding swing v_S at node S resulting from code-dependent switching. An entirely different technique, called "differential quad switching" [311, 338, 339], does not minimize this voltage swing (in fact, the swing itself increases), but rather aims at making the swing/glitch at S fully code-independent. The latter is true if S experiences an *equal* disturbance originated from the switches at *every* clock cycle (i.e. independently from the code/signal), instead of being disturbed only when the current's steering direction changes from one output terminal to another as a result of a new input code.

That is accomplished by adding a second pair of steering switches to the usual pair of switches as shown in the P-type current cell in Fig. 4.35. The four switches are driven by suitable logic ANDs between the regular control signals $c_{0+/-}$ and the output update clock Φ or its complement. The corresponding time-domain waveforms are sketched in Fig. 4.36. First of all, let us notice that in the quad scheme only one of the four switches is turned on for each conversion cycle. Moreover, on comparing the dual and quad switching schemes in Fig. 4.35 it is clear that in the dual switching scheme the steering direction of the current I_u is not changed by the addition of the new switches

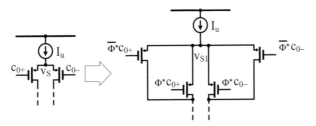

Figure 4.35 The differential quad switching current cell [338] (on the right-hand side) obtained as an extension of the traditional dual switching cell. Adapted from [311]

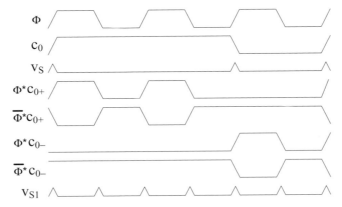

Figure 4.36 The waveforms associated with the dual and quad switching.

and it continues to be determined by c_0. However, whereas in the dual switching scheme the common source node is disturbed only when c_0 changes state, in the quad switching scheme even if c_0 does not change state during a given conversion cycle a pair out of the four switches will change state, hence creating a disturbance to the common source node. That is visible on the waveforms of Fig. 4.36 and, in particular, it is apparent when comparing v_S (for the dual switching scheme) with v_{S1} (for the quad switching scheme). The same effect can also be described by saying that the harmonic distortion resulting from the v_S glitches in the dual switching scheme is canceled out and replaced by tones at the sample frequency and its integer multiples. The latter are not signal-dependent and, besides, are far from the spectral content of V_{out}, which is located at very high frequency and also likely to be significantly filtered by the reconstruction filter following the DAC.

A few options for the respective frequencies for Φ and c_0 exist (e.g. "interleaved sampling," "double sampling"), providing some choice in terms of the power of the clock-dependent tones [311].

Besides the increase in strays at the common source node, the main drawback of the quad scheme is the increase of power consumption and complexity deriving from the addition of the extra switches and local logic to generate the additional control signals.

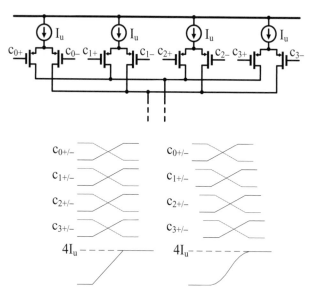

Figure 4.37 An example of the effect of timing skews on the output current of a thermometric DAC. On the right-hand side it is possible to see that a gradual skew in the state transition times can result in a rising current output with a different shape and longer transition time.

All and all, that is a reasonable price to pay given the significant distortion improvement provided by this approach [311].

4.3.4   Timing skews

Other important sources of dynamic nonlinearity becoming increasingly important for very high speed are the timing skews between the actual switching time instants of the multiple current-steering pairs (or quads). It has, in fact, been tacitly assumed so far that all current-steering pairs change their state (i.e. the direction of the steered currents from one output terminal to the other) at exactly the same time. However, inevitably, due to timing skews of different origins, each pair will in fact make its state transition at a slightly different time instant (in fact, to be completely accurate, even with slightly different transition times), leading to output distortion as exemplified in the simple DACs sketched in Figs. 4.37 and 4.38.

The sources of timing skew are multiple and include (see also Fig. 4.39)

- interconnect delays along the output lines carrying the individual current contributions to the output nodes;
- device mismatch in the steering switches;
- device mismatch in the switch drivers;
- interconnect delays along the clock lines triggering the state transitions for the latches;
- code-dependent load mismatch on the clock input for the latches;
- interconnect delays along the data busses to the current cells;
- load mismatch on the data busses to the current cells.

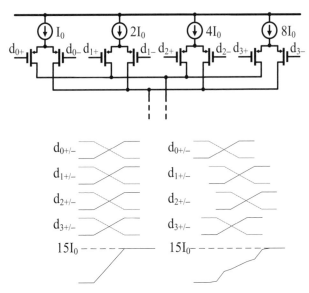

Figure 4.38 An example of the effect of timing skews on the output current on a binary weighted DAC. Timing skews can lead to even more dramatic distortion in this case than in the previous one.

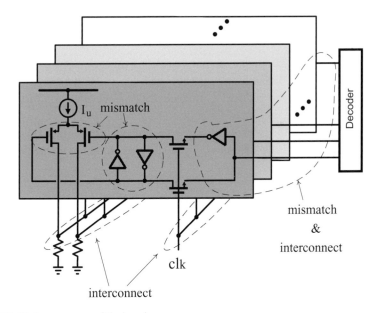

Figure 4.39 Various sources of timing skew.

Before discussing these issues in more detail it is important to make a few remarks that will provide a sense of the relevance of these matters.

First of all, Nyquist-rate current-steering DACs can have total physical dimensions ranging from about 0.5 mm to 1 mm even in advanced nanometer processes such as 65 nm or 90 nm CMOS processes [90, 321]. Therefore it is not surprising to have, for example, buffered clocks, data busses, and output current lines routed across the DAC and covering distances of this order of magnitude. With that in mind and considering rise and fall times of logic signals of the order of $t_r = 50$ ps, the corresponding wavelength of such signals is $\lambda \sim c \cdot 2t_r = 3 \cdot 10^8 \cdot 2 \cdot 50 \cdot 10^{-12} = 3$ mm, which is very comparable to the distances discussed and hence (1) interconnect delays simply cannot be ignored and (2) many long lines ought to be modeled as distributed lines [340].

Another important consideration is that propagation delays are not necessarily the real issue. The problem is rather the mismatch (that can be systematic, random, or signal/code-dependent) between delays which would otherwise be equal and fairly benign.

With these concepts in mind, let us consider the sources of timing skew previously enumerated and highlighted in Fig. 4.39.

Beginning with the output lines, each current will travel some distance from the steering switches to the output load. Hence, even if all steering switches in the array were to change their state at the same time, there will be differences in the timing at which each cell current reaches the load (the current waveform for cells located closer to the output nodes will experience shorter delay than will that for those located further from them). The contribution of this error source to the DAC's SFDR has been analyzed [341] and assessed to be

$$\text{SFDR} = \frac{4\left[1 + (2\omega_0\tau)^2\right]}{\tau_{max}\omega_0 \cos\left(\pi\omega_0/\omega_s\right)\sqrt{1 + \omega_0^2\tau^2}} \tag{4.24}$$

where $\omega_0 = 2\pi f_0$ is the output (angular) frequency, $\omega_s = 2\pi f_s$ is the output (angular) rate, τ is the output time constant,[15] and τ_{max} is the maximum time-constant variation. Here, the spread in delays between the cells is modeled as a variation in time constant; τ_{max} is such a maximum variation.

Similarly, the clock determining the time instant at which the cells change state may not reach all cells at the same time. The contribution of that to the SFDR is [341]

$$\text{SFDR} = \frac{4\sqrt{(1 + 4\omega_0^2\tau^2)/(1 + \omega_0^2\tau^2)}}{d_{max}\omega_0 \cos\left(\pi\omega_0/\omega_s\right)} \tag{4.25}$$

where d_{max} is the maximum delay difference between the cells.

The similarity between Eq. (4.24) and Eq. (4.25) is expectable since in either case the result is a delay mismatch between the cells' contributions to the total output. A number of observations on these results can be made. First of all, as expected, the SFDR degrades as the spreads in delays τ_{max} and d_{max} increase (clearly there is no corresponding contribution to distortion when all delays are the same, namely τ_{max} and d_{max} are zero). Moreover, the contributions to the SFDR depend on the ratio ω_0/ω_s between the output frequency f_0 and the sample rate f_s. The worst SFDR is seen for $f_0/f_s \sim 1/4$. For low

---

[15] The model assumed for the output settling is that the steering switches change state instantaneously, resulting in a current step into the output load, and that the resulting voltage response behaves as a single-pole $1/\tau = 1/(R_L C_L)$ system.

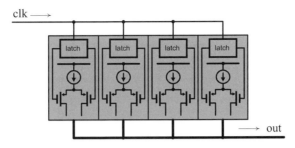

clk ⟶

out

Figure 4.40 Compensating the clock delays with the output delays.

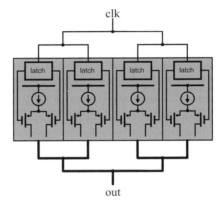

clk

out

Figure 4.41 Equalizing the clock delays as well as the output delays with clock T-tree structures.

output frequency ($\omega_0 \tau \ll 1$) Eqs. (4.24) and (4.25) simplify to

$$\text{SFDR} \simeq \frac{4}{\tau_{\max}\omega_0} \tag{4.26}$$

and

$$\text{SFDR} \simeq \frac{4}{d_{\max}\omega_0} \tag{4.27}$$

respectively, which means that the SFDR degrades with a -20 dB/dec rate over f_0 when the latter is low.

An implicit assumption in the output and clock timing skews just described is that these are systematic in the sense that they are due to uneven distances in the routing of the signals. Therefore, they can be (completely or partially) canceled out with proper routing techniques. For example, the similarity of Eq. (4.24) and Eq. (4.25) suggests that the two effects could be, at least partially, canceled out by providing a very delayed clock (larger d) to a cell that is very close to the output node (smaller τ) and vice versa as exemplified in the routing style depicted in Fig. 4.40 [311, 341]. A more complex but also more accurate approach consists of using clock trees as shown in Fig. 4.41 [311, 341] or Fig. 4.42. Similar considerations and equalization techniques can be applied to compensate for systematic timing skews in data lines. The availability of multiple metal levels

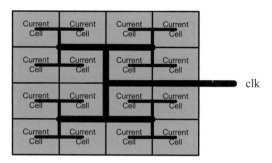

Figure 4.42 Equalizing line delays (in this example, the clock) with clock H-tree structures.

in the adopted process technology is of paramount importance for implementing such techniques and, where needed, to shield the multiple networks from one another.

A completely different source of timing skews is due to the mismatch in the steering switches and in the latches. Indeed, device mismatch in the steering switches introduces an input offset δc_0 into the current transfer characteristic of Eq. (4.23). This random (assuming that the steering switches have no systematic mismatch) voltage offset causes the steering of the current to begin earlier/later than desired when c_0 changes state. The issue is minimized if the transition of c_0 from one state to the other is very fast. The same can be said for the latches driving the steering switches: mismatches, particularly in the input devices of the latch (because the contribution of the other devices in the latch is reduced by the voltage gain between the input and the device at hand), result in input offsets and hence timing lead/lag. Once again, the faster the state change, the smaller the timing skew. Small switches offering low capacitive load to strong latches often lead to optimal solutions [334, 335].

Finally, as noted in previous sections, the output voltage $V_{\mathrm{outp/n}}$ couples to the common source node S through the steering switches. As a result, the "trip point" of the steering switches is once again disturbed (similarly to the just discussed case of mismatch between the steering switches), resulting in a signal-dependent timing skew [342]. Minimizing this coupling as discussed in Section 4.3.2 and, again, designing for the fastest possible steering transition are ways to mitigate this problem.

Code-dependent delays can also emerge in otherwise perfectly geometrically equalized structures making use of clock trees and suchlike, when the loads connected to the trees become code-dependent. That is the case, for example, of the fast latch structure shown in Fig. 4.29. The clock line that connects a large number of latches of this type will see a load that varies depending on how many of these latches change state from time to time since the channel and the stray capacitances of the switches forcing the state change may, or might not, need to vary its stored charge. Third-order distortion typically results from that [63].

The solution then consists of equalizing the load to the clock line as implemented in the modified latch shown in Fig. 4.43 [343, 344]. In this circuit, the simple latch of Fig. 4.29 has been augmented by an additional pair of switches (M_3 and M_4) and a set of inverters, which force the extra switches M_3 and M_4 to have a state that is always

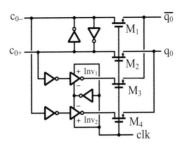

Figure 4.43 A clock-load-compensated latch [343, 344]. © 2007 IEEE.

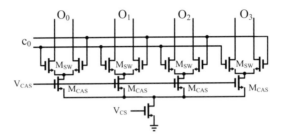

Figure 4.44 A universal current cell.

complementary to that of M_1 and M_2, hence making the load on the clock line higher but code-independent.

The last source of timing skews we will discuss in this section is due to large systematic differences in load between the segments of a DAC and within binary weighted sections of it. For example, in the segmented DAC depicted in Fig. 4.4, all transistors, including the steering switches, are sized proportionally to the currents they carry. This means that, to insure that all currents are truly steered concurrently (at least systematically), all the switches' drivers need also to be sized proportionally. When the number of bits in the binary weighted segment is relatively high this can become impractical and ways to re-balance the drivers or the loads need to be found.

A somewhat sloppy way consists of adding dummy loads to "faster nodes" to equalize the corresponding time constants. For example, in the scheme of Fig. 4.4, let us consider the weighted cells controlled by d_1 and d_2. Just to give an idea, these could be driven by the very same latch if the gates of the steering switches for the cell controlled by d_1 were also loaded by additional dummy PMOS transistors equal to the steering switches. Alternatively, the latch driving d_2 could be implemented by a suitable parallel connection of two latches identical to the single latch driving d_1.

A smarter solution, which also equalizes the time constants associated with internal nodes of the current cells, consists of creating a "universal current cell" used throughout the DAC array to implement all contributions, unary or weighted [345, 346]. For example, let us consider the cell shown in Fig. 4.44. A cascode-based splitter creates four identical currents equal to $I_u/4$. All of the steering switches are connected to the single control signal c_0. This single and modular cell can be used to create unary or binary weighted

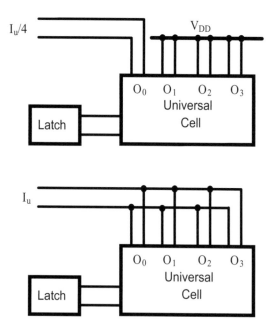

Figure 4.45 A universal cell connected to generate a current $I_u/4$ (top) or a current I_u (bottom).

currents such as I_u, $I_u/2$, and $I_u/4$ simply by properly shorting together the outputs O_0–O_3. An example is shown in Fig. 4.45. As stated earlier, however, regardless of how much current is steered to the output, the load on the latch is constant and all the time constants of the internal nodes are matched. Creating a layout with this cell is also very modular and straightforward since the amount of current steered can be hardwired simply by a via connection (or a missing via) between selected outputs O_0–O_3 and/or the DAC output rails ($V_{\text{outp}/n}$) or V_{DD}.

A similar alternative to that is the so-called "pseudo-segmentation" [347, 348]. This consists of implementing both the weighted and the unary cells as an "aggregation" of identical unity cells similarly to what was explained in Section 4.2.1 (see Fig. 4.7). This is an attempt at extending the performance advantages in linearity and scalability of classic thermometric DACs to arbitrarily segmented DACs. At the same time, the digital complexity of thermometric DACs (involving binary-to-thermometric decoders, their internal propagation delays etc.) is avoided by proper hardwiring of the involved cells as shown in the "pseudo-decoder" exemplified in the case of a 3-bit binary weighted DAC shown in Fig. 4.46.

4.3.5　Other causes of dynamic performance degradation

This section concludes the discussion on the causes of dynamic performance degradation and some of the techniques to mitigate their impact. There are a few more items to which the DAC designer may need to pay attention in order to be able to hit the target dynamic

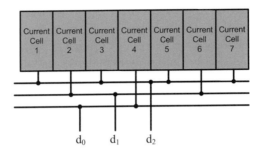

Figure 4.46 A pseudo-decoder for a 3-bit binary weighted DAC.

performance. Specifically, these include undesired coupling of digital noise into sensitive analog nodes and clock jitter.

Let us first concentrate on the coupling of digital noise generated by the multitude of digital circuits and signals existing in a DAC chip. Since the digital activity is likely going to be driven, in one way or another, by the DAC input codes, it will result in output distortion. This can manifest itself either as clearly visible harmonics of the intended output, or as mixing products of the output, its harmonics, and/or the clock frequency (or frequencies related to that), or as a broad range of spurious tones that, although harmonically related to the output in some form, have a somewhat "colored-noise" nature when observed with a spectrum analyzer.

Digital noise (or, better, digital disturbance) can couple into analog circuits in a variety of ways, including

- through physical proximity in the placement of "offending" digital circuits and "victim" analog circuits;
- intersection between conductors of digital signals and of analog signals;
- lack of proper isolation between digital and analog supplies.

Proximity can be the result of a strategy to minimize strays and area, and possibly improve matching. Intersection of wires can be inevitable due to space and routing restrictions. However, both of these also lead to undesired capacitive and inductive coupling.

This can happen in layout, for example when metal lines are close and/or parallel for sufficient length to build enough fringing or overlap capacitance or sufficient magnetic coupling. It can also happen through the substrate, particularly if that is epitaxial and, hence, has low resistivity along its surface. It can also happen in the package through bonding wires coupling or in the laminate of BGAs.

Sophisticated layout and package-extraction CAD tools exist, but, unfortunately, at present, all of the above mechanisms are still extremely hard to capture, properly model, and simulate, especially considering the level of accuracy required for high-performance converters. Besides, extracted netlists, due to the distributed nature of the strays, can be extremely large and therefore challenging both to the ability of simulators and computer servers to handle them, and to the ability of designers to actually gain insight into where the most important source of trouble may be coming from.

The skilled designer often needs to pair engineering acumen and intuition with a decent level of CAD know-how to efficiently anticipate, narrow down, isolate, and then reproduce with test cases and controlled simulation experiments the origins of the coupling and then find ways to mitigate them without disrupting the development schedule in the process. Regarding ways to mitigate the coupling, besides increasing the physical distance (often dramatically reducing the problems due to the exponential dependence of many of such strays on physical sizes) and/or changing the orientation of the coupled structures, the introduction of various types of interposed shields (tied to low-impedance signal grounds or sometimes left "floating") can "break" electric and magnetic field lines into controlled return paths for stray currents. Also, local substrate sneaky paths can be selectively altered by careful introduction of substrate contacts (again, tied to proper low-impedance AC grounds), by the use of insulating trenches, and/or by the introduction of buried doped wells (deep n-wells, triple wells and similar other options are often available in analog-friendly flavors of many CMOS processes).

Finally, in addition to the type of mechanisms described above, undesired AC coupling between signals and supplies or between otherwise isolated supply domains can also arise in the input/output (I/O) "ring" through voltage clamps and ESD protection circuits. These are all meant to monitor and limit mutual potential differences between supplies and signals. However they also come with many strays that can be non-negligible and, for sufficiently high frequency, provide plenty of detrimental coupling. Because of that the skilled designer is advised not to overly delegate to the ESD expert, who, perhaps due to excessive zeal or conservative ESD design, may unintentionally hand over an easily overseen source of trouble at the last minute.

Finally, sampling clock jitter is at present one of the toughest road blocks to high-dynamic-range performance when dealing with very-high-frequency signals. Just like for ADCs, the phase noise of the sampling clock is mixed with the output of a DAC and therefore can become the limiting factor for its noise performance. The dependence and manifestation of that is essentially the same as in ADCs and, therefore, will not be discussed much further here. It is, however, interesting to remark that, although the final results are essentially the same, the underlying mechanisms are not. In ADCs, aperture errors lead one to randomly sample the continuous-time input a bit too early/late and, hence, capture a signal that has a slightly different amplitude than what it would have had had it been sampled at the correct time. In DACs, instead, since the output is, in fact, a continuous-time signal, sampling clock phase noise leads to an output level that can randomly last a bit too long or not long enough compared with what would have been the case without phase noise. This also means that the charge delivered by the output current to the load can be too much/too little from period to period. In other words, the error isn't an amplitude error (as in ADCs) but rather a random timing-skew error. Once again, it is intuitive how that can become very important for rapidly varying outputs for which even small timing errors can be visible in comparison with the rate of change of the output signal. On the other hand, it won't be very visible on slowly varying signals (perfectly flat DC signals in the extreme case), for which the output step between one sample and the next one is minimal.

With that in mind, it is also quite intuitive why non-return-to-zero (NRZ) DACs have considerably less sensitivity to clock jitter than do return-to-zero (RTZ) DACs. NRZ DACs have only one clock edge per conversion, and output-level variations go only from the present output to the next one. On the other hand, RTZ DACs have two clock edges per conversion period (hence the phase noise is twice as much as in the NRZ case) and the jittery edges of the output can be quite large because the output returns all the way to (and from) zero in every period.

Finally, dual RTZ DACs [349], namely DACs whose output is the combination of two RTZ DAC outputs with the same output level but phase shifted by half a period, have the same jitter performance and output power as an NRZ DAC but with distortion performance comparable to (better than) that of an RTZ DAC [350].

4.4   Layout and floorplanning

The most effective way to keep control of strays[16] and, therefore, minimize the inevitable performance degradation when moving from a schematic design to an actual physical implementation is to start on the right foot by adopting a proper layout and floorplanning strategy.

There are generally two approaches to floor planning for current-steering DACs: a linear style and a matrix style (also referred to as Manhattan style). The linear style [90, 298, 311, 334, 351] consists of laying out the current cells, switches, latches, and some of the logic circuitry into identical slim and long slices and then abutting them in a straight line as depicted on the left-hand side of Fig. 4.47.

In the Manhattan style, on the other hand, the individual identical cells are shaped in a more square shape and are then arranged into a two-dimensional matrix as depicted on the right-hand side of Fig. 4.47 [306, 352].

More functional detail on the linear arrangement is shown in the diagram of Fig. 4.48. Owing to its relative simplicity, the linear style provides a straightforward way to closely abut devices that are meant to match one another as evidenced in Fig. 4.48, where it is possible to distinguish a separate area for the current-source devices M_{CS}, one for the combination of the cascode M_{CAS} and steering switches M_{SW}, and one for the drivers/latches. The physical separation among these sections also minimizes undesired coupling between them; such as coupling between the very active digital signals in the latches, decoders etc. and the quiet current-source devices.

Common signals such as supply lines, biasing, output busses etc. are routed through the arrays as appropriate. Clock-tree structures such as those described in the previous sections can be adopted for actual clock signals and outputs, but also for other types of global nets such as biasing lines and power supplies [337]. The reason for the latter is that, if the associated stray resistances and capacitances cannot be sufficiently minimized due to their aggressive corresponding requirements, they can at least be equalized among the cells to better mitigate the corresponding detrimental effects discussed in the previous

---

[16] Or to jeopardize an otherwise fine circuit design.

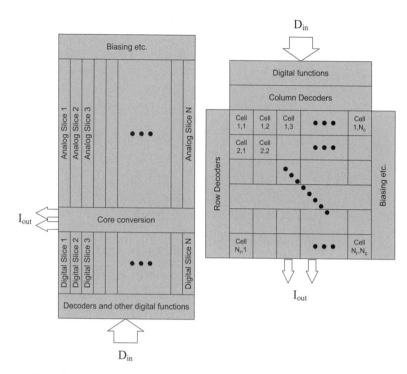

Figure 4.47 DAC floorplans: (left) linear style and (right) matrix/Manhattan style. Adapted from [348] and [312].

sections. Once again, the availability of multiple metal levels is a powerful enabler of such techniques insofar as it allows layering and, to some extent, shielding and decoupling among multiple global nets over the large array of cells. It is, in fact, also because of the increased availability of additional multiple metal levels in advanced CMOS processes that there has recently been an increase in popularity of this floorplanning style over the more traditional Manhattan style.

A more detailed diagram showing an example of the Manhattan floorplan is depicted in Fig. 4.49. Although in this case the interconnects between the transistors constituting the current cells are potentially more compact than in the linear floorplan, on the other hand, a considerable number of long critical lines must be run through the entire array. In fact, both digital lines with a fair amount of activity and critical analog lines, such as the bias lines and the output rails, run across the entire array and cross to one another in multiple places. That can cause significant coupling between the respective signals and can ultimately result in visible harmonic distortion and a broad range of spurious tones in the output spectrum.

4.5   Dynamic element matching in segmented Nyquist-rate DACs

Dynamic element matching (DEM) is a well-known class of digital techniques allowing one to improve the distortion performance of DACs at the expense of an increase in

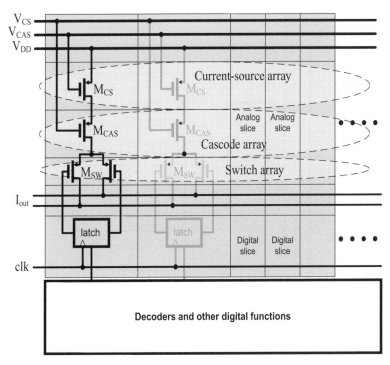

Figure 4.48 The linear floorplan. A current-source array, a cascode array, a switch and a latch array are shown together with the repeated modular slices.

the NSD by pseudo-randomly scrambling the usage pattern of its constituting cells [68]. Some DEM algorithms, called mismatch-scrambling algorithms, turn the distortion tones into white noise and therefore are better suited to wideband, Nyquist-rate converters. Other DEM algorithms, called mismatch-shaping algorithms, turn the distortion spurs into a frequency-shaped noise and, therefore, are a good fit for oversampled converters.

DEM techniques have been widely and successfully used in DACs built from a relatively low number of 1 b sub-DAC cells such as in $\Delta\Sigma$ ADCs [81], $\Delta\Sigma$ DACs [353, 354], and sub-DACs of pipelined ADCs [281, 355]. These are very well-established techniques, the corresponding literature is vast, and a discussion on that is well beyond the scope of this book.

However, very recently, examples of DEMs in current-steering Nyquist-rate DACs with many constituting elements (the type of DACs discussed in this chapter) have been reported. Providing a glance at these advanced and somewhat specialized DEMs is the topic of this section.

A number of challenges have limited wider use of DEMs in Nyquist-rate DACs. First of all, the effectiveness of element scrambling relies on time-averaging the undesired output deviations (due to mismatches and leading to nonlinearity). If, after scrambling, on average, the output levels are correct then the deviations from a linear response are turned into a randomized addition, namely into a noise-like waveform superimposed on the desired output. For the averaging to effectively take place the output waveform frequency needs to be considerably smaller than the frequency at which the elements

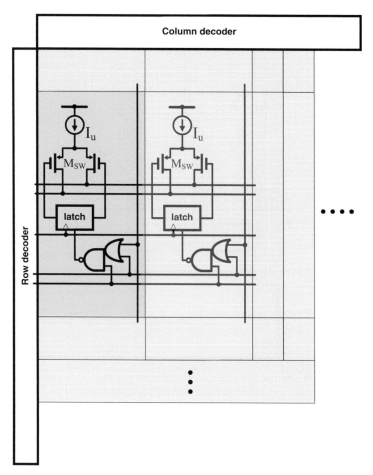

Figure 4.49 A matrix floorplan. A unity cell contains here both analog and digital blocks. The state of the switches is determined by local logic sensing the state of the row and column code. Even though this is a very simplified diagram (e.g. the current-source bias lines aren't even sketched) the multiple intersections between analog signals and digital signals are apparent.

are scrambled. It is clear that this is easily satisfied in oversampled converters, where the signal band is, by definition, much smaller than the sample rate. That is not the case for Nyquist-rate DACs unless the output signal is at a much smaller frequency than the Nyquist frequency (i.e. essentially falling again in the case of oversampled signals).

Another issue has long been associated with the implementation of the DEM scrambler placed between the input data stream $\mathbf{D}_{in}$ and the actual DAC. Specifically, conventional implementations of DEM scramblers assume that all the DAC elements being scrambled are nominally identical (i.e. that the scrambled DAC is a fully thermometric one). These algorithms, in fact, take advantage of the natural "redundancy" present in thermometric DACs: if all unit elements are indeed the same then the desired final summed current can be equally well obtained through multiple selections of the elements as long as the total number of selected elements is constant. So the DEM will always pick a different combination of unit (actually mismatched) elements, hence scrambling the distortion patterns into uncorrelated noise-like spurs.[17] However, since DEM scramblers make use of cascaded bit-shuffling blocks (also known as "butterfly networks") arranged into tree-like structures, the size of the scrambler grows exponentially with the number of bits, quickly becoming impractical for DACs with several output levels.

Because of all that, some early attempts at introducing DEM scramblers into Nyquist-rate current-steering DACs were limited to partial sections of thermometric segments [298, 301, 351] and, therefore, did not always provide their full potential benefit [351]. In fact, since the addition of the DEM scrambler could introduce more (possibly code-dependent) digital activity into what could be an already potentially noisy environment, the relative benefit in terms of DC and low-frequency distortion improvement may end up being unintentionally traded for a further degradation of the high-frequency behavior. A way to effectively mitigate this problem in mismatch-shaping DEMs has been reported in [357, 358], where the key idea is keeping the number of switching events per period constant so that the disturbances as a whole are no longer directly code-dependent.

The same general idea has also been applied (this time in a mismatch-scrambling case) to the zeroing cycle of the output of a return-to-zero (RZ) DAC in [321], where the output is nulled by steering the currents of half of the elements to the positive output and half to the negative output; however, the order of selection of the current elements is always different and randomized and, hence, static as well as dynamic mismatches are scrambled into noise.

As pointed out above, one of the obstacles to introducing DEMs for Nyquist-rate DACs with multiple current cells has been that the DEM encoder's complexity and the number of shuffled unit elements grow exponentially with the number of bits of DAC resolution [68]. Furthermore, while the thermometric DAC structure provides the previously mentioned "redundancy" (i.e. in the absence of mismatch, the particular selection of the units is irrelevant) exploited by traditional DEMs, in a binary weighted DAC the elements constituting the array are not interchangeable since each has a different weight.

For intuitive reasons, some of these problems resemble those of implementing a high-resolution DAC (say, more than 8–10 bits) with a fully thermometric approach. Then, in analogy to the segmentation approach followed for high-resolution DACs, "segmented DEMs" have been introduced first in $\Delta\Sigma$ converters [350, 359, 360, 361] and, more recently, in Nyquist-rate DACs [356, 362, 363, 364] with segmentation.

While in [350, 360, 361] the different segments have different weights but each is still individually thermometric and hence independent DEM scramblers are used for each of them (hence the segmentation boundary mismatch needs to be separately addressed), in [356, 359, 362, 363, 364] the segmented DEM scrambler controls the full segmented DAC. One of the keys to the latter approaches consists of intentionally introducing the needed "redundancy" among the binary weighted elements by adding more binary weighted cells to the existing ones. The redundant binary weighted elements are then

---

[17] The requirement for the scrambler is that (1) each choice of elements is statistically independent from the choices made in other sample periods and (2) all of the possible sets of chosen elements have an equal probability of being chosen [356].

used in an exclusive way (i.e. if two cells of equal weight K_i now replace what was originally a single cell of the same weight K_i, only one of the two cells is actually used to steer current to the output at a time, while the other one dumps its current to a ground node) and the selected cell changes from cycle to cycle as directed by the DEM scrambler.

Note that, on introducing redundant cells/elements, the actual DAC resolution (or, in other words, the number of output levels) increases. However, as explained above, the actual *usable* input range of the DAC does not increase. For example, in [356], a 14 b DAC with a 4–10 segmentation is described. Without the segmented DEM the DAC would have had $2^4 \cdot 2^{10} - 1 = 16\,383$ (i.e. $2^{14} - 1$) output levels. To introduce the required redundancy and a segmented DEM scrambler, ten additional binary weighted cells are added to the binary weighted LSB segment. The resulting structure could then provide $2^4 \cdot (2^{10}) + (2^{10}) - 1 = 17\,407$ levels; however, as explained above, only $2^{14} - 1$ are actually usable since the redundant sources are only mutually exclusively steered to the output. All that is a fundamental limitation of the segmented DEM approach.

The structure of the segmented DEM scrambler is still a tree structure. However, the segmentation limits the complexity of the scrambler considerably. While the sub-network of switching blocks controlling the thermometric section still grows exponentially with the number of corresponding segment bits, on the other hand, the remainder of the scrambler essentially grows linearly with the size of the binary weighted segment [356, 364].

Finally, a very important point associated with these DEM methods is that they scramble not simply the static mismatches (i.e. the static currents in the cells) but also the entire waveforms associated with the constituting cells. Therefore nonlinearities associated with timing mismatches between these cells are also scrambled. What the DEM cannot address, since it does not stem from selection order or mutual mismatch, is, for example, the finite output impedance of the cells [356, 364].

Confirming the above considerations, the SFDR performance plots (versus the output tone frequency f_o) of the scrambled DAC reported in [356, 362, 363] are remarkably flat and begin dropping only when the impedance limitation starts limiting them. Once again, just like for every DEM algorithm, since the spurious power initially concentrated on the harmonics is scrambled into noise-like spurs, the NSD after scrambling is degraded compared with the NSD before scrambling. Likewise, the SNDR will, in the best case, stay the same or, possibly, degrade due to the additional switching introduced by the DEM.

4.6   Signal processing techniques

Besides the circuit techniques discussed in the previous sections, various signal processing (SP) techniques are increasingly associated with high-speed DACs, being used to maximize their performance in specific circumstances and applications. In part, the introduction of such techniques has been the natural result of the access to higher integration of digital functions in advanced CMOS processes. For example, many commercial DACs include significant on-chip digital pre-processing functions such as interpolation/filtering and digital frequency up-conversion ahead of the actual conversion into an analog output. On the other hand, other signal processing techniques are either partially

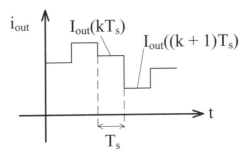

Figure 4.50 In an NRZ DAC the output swings from one converted level to the next one at every period T_s.

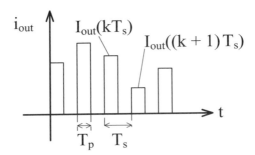

Figure 4.51 In an RTZ DAC the output swings from one converted level to zero before heading toward the next converted level at every period T_s. The output is only non-zero for a time T_p.

or entirely implemented in analog form to enhance or suppress specific features of the analog behavior of the converter itself, as will be shown in some of the following examples. A quick overview of these techniques is the topic of this section.

To begin with, it is useful to recall some known relations describing the time- and frequency-domain behavior of the output of an ideal DAC. In most of the previous sections we have tacitly assumed that we are considering DACs with a continuous-time output $i_{out}(t)$ swinging from $I_{out}(kT_s)$ to $I_{out}((k+1)T_s)$ as a result of input codes $\mathbf{D}_{in}(k)$ and $\mathbf{D}_{in}(k+1)$, respectively, at subsequent sample times ($T_s = 1/f_s$ being the sample/output time period) as shown in Fig. 4.50. That is what is commonly referred to as a *non-return-to-zero* (NRZ) DAC.

However, in general, it may be desirable (as explained later) to design a DAC whose output at time $t = kT_s$ stays at the converted value $I_{out}(kT_s)$ for only part of the period T_s while it goes to zero during the remainder of T_s, before it goes to the next converted value $I_{out}((k+1)T_s)$ at $t = (k+1)T_s$ and so on, as depicted in Fig. 4.51. A DAC with such behavior is called a *return-to-zero* (RTZ) DAC.

In general, a well-known time-domain representation for a DAC output such as the one shown in Fig. 4.51 is

$$i_{out}(t) = \left[\sum_{n=-\infty}^{+\infty} I_{out}(nT_s)\delta(t - nT_s) \right] \otimes \mathrm{rect}\left(\frac{t}{T_p} \right) \tag{4.28}$$

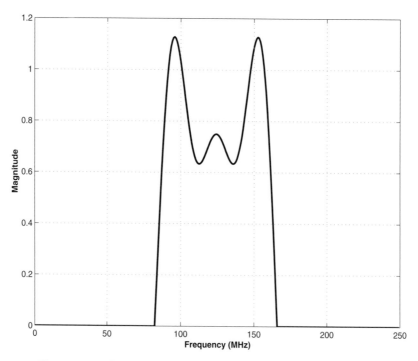

Figure 4.52 The spectrum of an example of a narrowband digital signal $\mathbf{I}_{out}(f)$.

where $T_p, 0 < T_p \le T_s$, is the time during which the output is held at $I_{out}(nT_s)$, $\otimes$ is the convolution operator, $\delta(t)$ is the Dirac delta function, and rect(t) is the rectangular function:

$$\text{rect}(t) = \begin{cases} 0 & \text{if } |t| > 0.5 \\ 0.5 & \text{if } |t| = 0.5 \\ 1 & \text{if } |t| < 0.5 \end{cases} \tag{4.29}$$

The Fourier transform of $i_{out}(t)$ in Eq. (4.28) is

$$\mathbf{i}_{out}(f) = \mathbf{I}_{out}(f) \left[\sum_{n=-\infty}^{+\infty} \delta(f - n \cdot f_s) \right] \frac{T_p}{T_s} e^{-j\pi f T_p} \text{sinc}(f T_p) \tag{4.30}$$

where $\text{sinc}(f T_p)$ is the *normalized sinc function*:

$$\text{sinc}(f T_p) = \frac{\sin(\pi f T_p)}{\pi f T_p} \tag{4.31}$$

For example, in the case of an NRZ DAC ($T_p = T_s$), the magnitude plots for a possible narrowband signal $\mathbf{I}_{out}(f)$ and the corresponding output $\mathbf{i}_{out}(f)$ are shown in Figs. 4.52 and 4.53, respectively. In the latter, it can be seen that, in addition to aliasing, the output spectrum $\mathbf{i}_{out}$ suffers from magnitude distortion due to the sinc function. This results in a magnitude suppression that is greater in higher Nyquist bands.

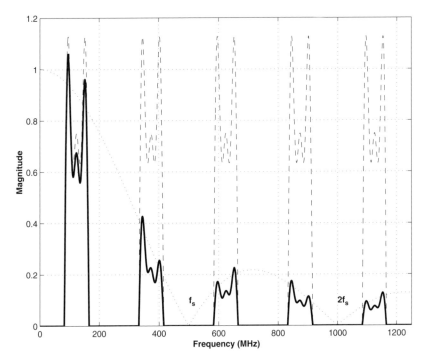

Figure 4.53 The spectrum of the converted NRZ DAC output $i_{out}(f)$ (solid line). For convenience, the undistorted aliases of $\mathbf{I}_{out}(f)$ are shown as dashed lines and the envelope of the $|(T_p/T_s) \cdot \text{sinc}(fT_p)|$ function is reported as a dotted line. Note also the nulls of the sinc function at kf_s. In this example $f_s = 500\,\text{MHz}$.

In a regular NRZ DAC appreciable distortion can even be detected within the first Nyquist band $(0 < f < f_s/2)$ since, for example, at the Nyquist frequency $(f = f_s/2)$ we have $\text{sinc}(f \cdot T_p) = \text{sinc}(f_s/2 \cdot T_s) = \text{sinc}(0.5) = \sin(\pi/2)/(\pi/2) = 2/\pi \simeq 0.64$ (namely $-3.9\,\text{dB}$). If necessary, in the first Nyquist band, the latter can be corrected by means of digital pre-distortion/emphasis [365].

If the signal $\mathbf{I}_{out}(f)$ of the example of Fig. 4.52 is converted using an RTZ DAC with $T_p = T_s/2$ the resulting output spectrum will change as shown in Fig. 4.54. If, instead, an RTZ DAC with $T_p = T_s/3$ is used, the output spectrum will be that shown in Fig. 4.55.

A plot of the sinc function for different T_p/T_s ratios is shown in Fig. 4.56 and presents another important consideration, namely that, although an RTZ DAC suffers less from sinc distortion (only in the sense that the aliases are a bit more equal to each other), the power of the output $i_{out}(t)$ decreases (which is intuitive, since there is actually less non-zero output during T_s) with understandable detrimental impact, for example, on the SNR.

Because of that, a brute-force way to restore the overall output signal power is to increase the output amplitude by increasing the full-scale output current I_{FS}. The latter may inevitably impact the entire design of the DAC, its voltage headroom, and its distortion performance.

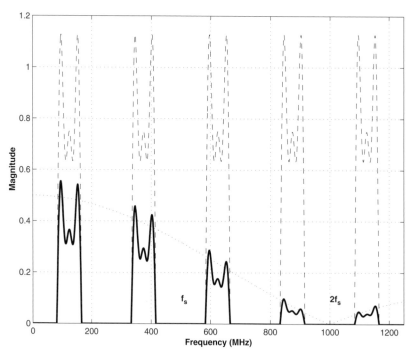

Figure 4.54 The spectrum of the converted RTZ DAC output $\mathbf{i}_{\text{out}}(f)$ (solid line) for $T_{\text{p}} = T_{\text{s}}/2$. Note the nulls of the sinc function at $2kf_{\text{s}}$. In this example $f_{\text{s}} = 500\,\text{MHz}$.

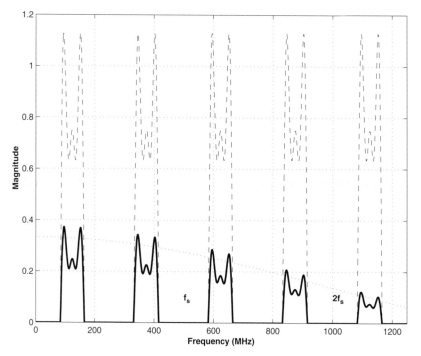

Figure 4.55 The spectrum of the converted RTZ DAC output $\mathbf{i}_{\text{out}}(f)$ (solid line) for $T_{\text{p}} = T_{\text{s}}/3$. Note the nulls of the sinc function at $3kf_{\text{s}}$. In this example $f_{\text{s}} = 500\,\text{MHz}$.

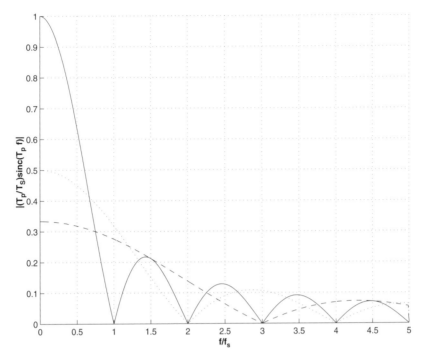

Figure 4.56 The envelope of the sinc distortion for various sizes of the DAC output pulse. The NRZ ($T_p = T_s$) is shown as a solid line. The RTZ cases for $T_p = T_s/2$ and $T_p = T_s/3$ are shown as dotted and dashed lines, respectively.

Despite these issues, besides a reduced sinc distortion, there are valuable reasons why an RTZ DAC may be a suitable choice in some cases. First of all, when an NRZ DAC swings from one output level to the next one every T_s seconds, its output might not have sufficient time to settle to the desired final value within T_s. As result of that, when the subsequent digital input is converted, the corresponding new analog output will also depend on the prior initial condition (which is the prior unsettled condition). If the output settling follows the laws of a linear system (namely, this could be modeled as a perfectly linear DAC followed by a linear dynamic system, such as a linear filter) then no distortion is introduced because of that. However, if the settling is nonlinear then this initial condition dependence results in distortion. A related example of this type of nonlinear settling is when the DAC output exhibits a different settling behavior when rising from a lower output level to a higher output level than when falling from the same upper level to the same lower level. That issue can be visible, for example, in single-bit DACs ($\Delta\Sigma$ DACs) for audio applications. These phenomena are known as *nonlinear inter-symbol interference* (nonlinear ISI) [366].

Assuming that it is possible to re-set the output to zero at every cycle, as in an RTZ DAC, the memory of the prior output is erased before a new one is produced. This eliminates or dramatically reduces ISI and its associated distortion.

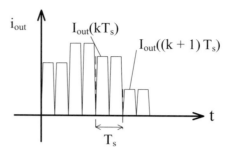

Figure 4.57 In a dual RTZ DAC the output is effectively the combination of two RTZ outputs with the same value but shifted by half a period.

Second, as we have seen in previous sections, many causes of dynamic linearity problems in DACs occur when the steering switches change state (e.g. timing mismatches), resulting in a transient nonlinear settling of the output that eventually dies out. If these "glitches" are sufficiently short-lived (at least, sufficiently short relative to T_s) then one way to "mask them" to the output is to steer the output current to a ground node while the glitch is happening and then re-route the current to the actual output node after its effect has died out. Hence the RTZ output appears "glitch"-free, namely, substantially free of distortion spurs [297].

Once again, the dual RTZ technique [349] previously cited in Section 4.3.5 (see an example of a dual RTZ output in Fig. 4.57) is characterized by a sinc distortion equal to that of an NRZ DAC, and also has the correspondingly stronger output power and lower clock jitter sensitivity, while at the same time addressing some of the cited distortion mechanisms as in an RTZ DAC. This technique, while very successful in audio DACs [350], has not yet been sufficiently proven in applications for which both very high speed (output frequencies in excess of hundreds of MHz) and very high dynamic range (harmonic distortion greater than seventy or more dB) are needed simultaneously, such as in wireless infrastructure transmitters.

An interesting phenomenon happens if, instead of repeating the same output level twice within the same period as in a dual RTZ DAC, the output level during the second half of the output period is instead made equal in amplitude but opposite in sign, as shown in Fig. 4.58 [367, 368]. So, applying Eq. (4.30) for $T_p = T_s/2 = 1/(2f_s)$, if the Fourier transform of the RTZ output is given by

$$\mathbf{i}_a(f) = \mathbf{I}_{\text{out}}(f) \left[\sum_{n=-\infty}^{+\infty} \delta(f - n \cdot f_s) \right] \frac{1}{2} e^{-j\pi f/2f_s} \operatorname{sinc}[f/(2f_s)] \qquad (4.32)$$

then, the Fourier transform of the signal shown in Fig. 4.58, called "mixed mode," is given by

$$\mathbf{i}_{\text{out}}(f) = \mathbf{i}_a(f)(1 - e^{-j\pi f/f_s})$$

$$= \mathbf{i}_a(f)(1 + e^{-j\pi} e^{-j\pi f/f_s})$$

$$= \mathbf{i}_a(f)(1 + e^{-j\pi(1+f/f_s)})$$

$$= 2\mathbf{i}_a(f) e^{-j\frac{\pi}{2}(1+f/f_s)} \cos\left[\frac{\pi}{2}\left(1 + \frac{f}{f_s}\right) \right] \qquad (4.33)$$

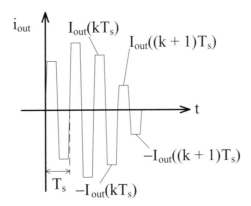

Figure 4.58 In a mixed-mode DAC the output can be seen as the combination of two RTZ outputs with equal amplitude and opposite sign, shifted by half a period.

Hence, in this case, the sinc distortion is replaced by a different frequency-shaping function, resulting in an attenuation of the output signal power in the first Nyquist band while the power of the alias in the second Nyquist band is instead increased as shown in the upper plot of Fig. 4.59. This is very advantageous if the intended synthesized signal is actually the image in the second Nyquist band (as opposed to the one in the first Nyquist band). That image has visibly a higher power than in a traditional NRZ or RTZ DAC, and the other aliases are eventually further suppressed by using a passband filter (selecting only the image in the second Nyquist band) instead of a traditional lowpass reconstruction filter.

The implementation of the mixed-mode waveform shown in Fig. 4.58 could be done by proper convolution, in the digital domain, between the series $\mathbf{D}_{in}$ and a suitable series of 1s and -1s before the actual digital-to-analog conversion. However, it is much more easily obtained using the differential quad switching already presented in Section 4.3.3 with intuitive modifications of the sequence of the state/sign of the control signals for the four switches [367, 368].

The above mixed-mode scheme is actually a particular case of a more general approach whereby the DAC output can be built as a proper combination of multiple input-data-controlled pulse trains (each one shifted in phase from the next) as described, for example, in [369]. The advantage of these signal processing approaches, once again, is that one is able to emphasize and de-emphasize specific images of the input spectrum [370] so that the DAC can actually directly synthesize a very-high-frequency output signal starting from a low-frequency input. In this way the DAC essentially embeds an up-conversion mixer, which, for example, can then be eliminated from the following analog signal chain in a transmitter application.

An alternate implementation of the mixed-mode technique employs two parallel DACs, both clocked at the output rate f_s and fed with the same input data series $\mathbf{D}_{in}$ at f_s. However, the two DACs are clocked with a phase shift of 180°, hence also creating two phase-shifted outputs i_{out1} and i_{out2}, which are then subtracted to obtain i_{out} as shown in Fig. 4.60.

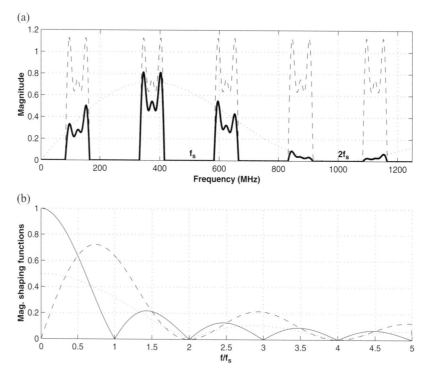

Figure 4.59 The frequency shaping for the mixed-mode DAC: (a) while the power of the alias in the second Nyquist band is increased, the other aliases are attenuated; (b) the envelopes of the magnitude of the sinc distortion for an NRZ DAC (solid) and an RTZ DAC with $T_p = T_s/2$ (dotted), and the shaping of the mixed-mode DAC (dashed). In this example $f_s = 500\,\text{MHz}$.

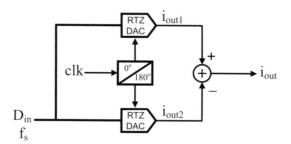

Figure 4.60 The block diagram for an alternate implementation of the mixed-mode DAC.

In a slight variation of this technique, presented in [371], the two DACs are NRZ DACs (instead of RTZ DACs as in the mixed-mode technique) and their outputs are summed (instead of being subtracted) as shown in the block diagram depicted in Fig. 4.61. As a result of that, the spectrum of the NRZ i_{out1} is

$$\mathbf{i}_{\text{out1}}(f) = \mathbf{I}_{\text{out}}(f) \left[\sum_{n=-\infty}^{+\infty} \delta(f - nf_s) \right] e^{-j\pi f/f_s} \operatorname{sinc}(f/f_s) \qquad (4.34)$$

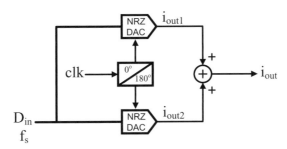

Figure 4.61 The block diagram for an image-cancelation parallel DAC [371]. © 2004 IEEE.

while the spectrum of the resulting output $i_{\text{out}} = i_{\text{out1}} + i_{\text{out2}}$ is

$$\mathbf{i}_{\text{out}}(f) = \mathbf{i}_{\text{out1}}(f)(1 + e^{-j\pi f/f_s})$$

$$= 2\mathbf{i}_{\text{out1}}(f)e^{-j\pi f/(2f_s)} \cos\left(\frac{\pi}{2}\frac{f}{f_s}\right)$$

$$= 2\mathbf{I}_{\text{out}}(f)\left[\sum_{n=-\infty}^{+\infty} \delta(f - nf_s)\right] e^{-j\pi 3 f/(2f_s)} \cos\left(\frac{\pi}{2}\frac{f}{f_s}\right) \operatorname{sinc}\left(\frac{f}{f_s}\right) \quad (4.35)$$

which, not surprisingly, closely resembles the mixed-mode equation (4.33) (although, again, the sinc distortion in i_{out1} is different from that of i_a due to the use of two NRZ DACs). Noting that $\cos\left(\frac{1}{2}\pi f/f_s\right) = 0$ for $f = f_s + 2mf_s$ and $m = 0, \pm 1, \pm 2, \ldots$, whereas it is $\cos\left(\frac{1}{2}\pi f/f_s\right) = (-1)^{f/(2f_s)}$ for $f = 2mf_s$ and $m = 0, \pm 1, \pm 2, \ldots$, Eq. (4.35) can be further rewritten as

$$\mathbf{i}_{\text{out}}(f) = \mathbf{I}_{\text{out}}(f)\left[\sum_{n=-\infty}^{+\infty} (-1)^n \delta(f - 2nf_s)\right] e^{-j\pi 3 f/(2f_s)} \operatorname{sinc}\left(\frac{f}{f_s}\right) \quad (4.36)$$

By observing the periodicity of the delta series (which determines the aliases of $\mathbf{I}_{\text{out}}(f)$) in Eq. (4.36) it can be seen that half of the original images of $\mathbf{i}_{\text{out1}}(f)$ and $\mathbf{i}_{\text{out2}}(f)$ have been canceled out. Specifically, $\mathbf{i}_{\text{out}}(f)$ has only the baseband image of $\mathbf{I}_{\text{out}}(f)$, its image in the third and fourth Nyquist bands, and on and on with this periodicity, as depicted in the lower spectrum of Fig. 4.62. Unfortunately, however, the sinc distortion on $\mathbf{i}_{\text{out}}(f)$ is still that of the NRZ DACs with output rate f_s.

Also, as expectable, in practice the image suppression might not be as complete as described above if, for example, the timing skew between the two DACs isn't as perfect as desired. So, if ΔT_0 is the delay between the clocks driving the two DACs, ideally it should be $\Delta T_0/T_s = 1/2$. Referring to the image on the second Nyquist band, it is possible to define the *Nyquist-image suppression ratio* (NISR) as the ratio between the Nyquist-image amplitude without suppression ($\Delta T_0/T_s = 0$) and the Nyquist-image amplitude with incomplete suppression ($\Delta T_0/T_s \simeq 1/2$). It can be proven [371] that the

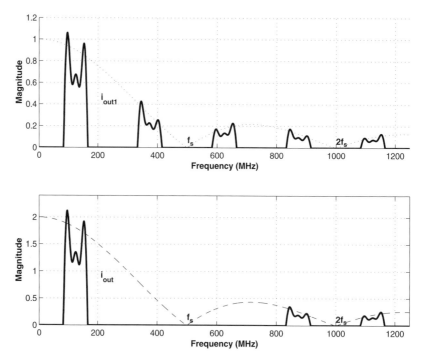

Figure 4.62 The frequency shaping for an image-cancelation parallel DAC. The output i_{out1} containing all aliases is depicted on the upper plot, while the final output i_{out} with the canceled second and third images is depicted on the lower plot. In this example $f_s = 500$ MHz. Adapted from [371].

NISR is given by

$$\text{NISR} \simeq \frac{1}{|\cos(\pi \, \Delta T_0 / T_s)|} \tag{4.37}$$

Therefore, as expected, the suppression is infinite for perfect time-shift $\Delta T_0 / T_s = 1/2$ but degrades rapidly as the timing-skew error increases.

Similarly, gain mismatch, offset mismatch, and other mismatches between the two parallel DACs will also degrade the quality of the image suppression (although, as is intuitive, the timing skew is the hardest mismatch to control). In spite of these similarities with time-interleaving (discussed in Section 3.4), this approach does not exactly represent a form of time-interleaving since the outputs of the two DACs are always present and summed; moreover, the output is shaped by $\text{sinc}(f/f_s)$ instead of $\text{sinc}(f/(2f_s))$ [371].

The issues associated with the above mismatches can be eliminated by moving the signal processing cancelation described by Eq. (4.36) into the digital domain and then following it with a single DAC. However, in this case, the cost to the single DAC is not negligible, either, since this DAC's output rate needs to double to $2f_s$ [371].

This brings us seamlessly to the last and most common signal processing technique described in this section, which can be used to minimize the sinc distortion and/or

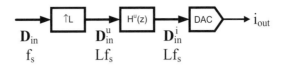

Figure 4.63 The block diagram for the cascade of an up-sampler, an interpolation filter, and a DAC.

simplify the design of the reconstruction filter by spacing away the output signal images. It consists of *interpolating* the input data series $\mathbf{D}_{in}$ sampled at f_s, namely another data series $\mathbf{D}_{in}^u$ at a higher sampling rate $L \cdot f_s$ is created from $\mathbf{D}_{in}$ using a digital *up-sampler* as follows [372]:

$$\mathbf{D}_{in}^u(n) = \begin{cases} \mathbf{D}_{in}(n/L) & \text{if } n = 0, \pm L, \pm 2L, \ldots \\ 0 & \text{otherwise} \end{cases} \tag{4.38}$$

where L is a positive integer greater than unity (often conveniently set as a power of 2). A block diagram of the processing functions is sketched in Fig. 4.63. As can be seen from Eq. (4.38), for each code $\mathbf{D}_{in}$ during T_s, a new set of samples $\mathbf{D}_{in}^u$ is created: the first one is equal in value to $\mathbf{D}_{in}$ and the other $L - 1$ samples are zeros (hence the term "zero stuffing" [373]). In the frequency domain this results in

$$\mathbf{D}_{in}^u(f) = \mathbf{D}_{in}(Lf) \tag{4.39}$$

namely, although the sample rate of $\mathbf{D}_{in}^u(f)$ has increased to Lf_s, its spectrum shows images of $\mathbf{D}_{in}(f)$ every f_s as shown in Fig. 4.64.

The introduced zeros can then be replaced by appropriate non-zero values by an interpolation process, which, in the ideal case, is achieved by passing $\mathbf{D}_{in}^u(n)$ through an ideal "brick-wall" lowpass filter with a frequency response

$$H^u(f) = \begin{cases} L, & |f| \leq 1/(2L) \\ 0, & 1/(2L) < |f| < 1/2 \end{cases} \tag{4.40}$$

Namely, after this filtering, only the first image is left and all of the other $L - 1$ unwanted images are suppressed [372, 373]. The resulting output $\mathbf{D}_{in}^i(f) = H^u(f)\mathbf{D}_{in}^u(f)$ is the *interpolated* series and has the spectrum shown in Fig. 4.64, namely the frequency spacing of the images and the sinc distortion has broadened L times to a wider frequency range. Since the interpolated series $\mathbf{D}_{in}^i$ is then fed to the DAC, the main cost for this transformation is that the DAC now needs to have an L times higher output rate.

Also, in general, the interpolation filtering function of Eq. (4.40) could be replaced by another type of filter, such as a bandpass or a highpass filter, hence selecting higher-frequency images instead of the "baseband" image.

Most modern communication DACs have digital interpolation filters in front of the actual converter, allowing a lower input data rate to the chip, especially when narrowband signals are being converted, and then increasing the rate internally for $L = 1, 2, 4, 8, \ldots$

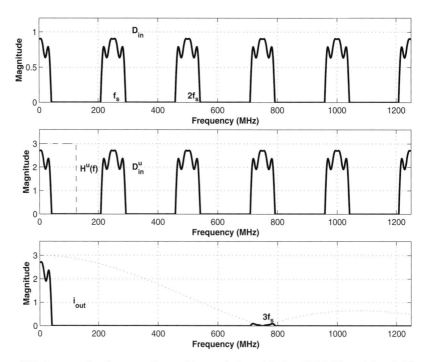

Figure 4.64 An example of up-sampling and interpolation with $f_s = 250\,\text{MHz}$ and $L = 3$. The original series $\mathbf{D}_{\text{in}}$ sampled at a rate f_s is shown in the upper plot. The up-sampled (zero-stuffed) series $\mathbf{D}_{\text{in}}^{\text{u}}$ is shown in the middle plot as a solid line. The transfer function $H^{\text{u}}(f)$ of the interpolation filter is shown overlapped as a dashed line for convenience. The NRZ DAC output at rate $3 f_s$ synthesizing the interpolated series $\mathbf{D}_{\text{in}}^{\text{i}}$ is shown in the bottom plot together with the magnitude envelope of the 3 sinc function as a dotted line.

This section ends with a note of caution. The use of the above-described techniques allows emphasis or de-emphasis of some of the images in order to mitigate sinc distortion and/or magnify a higher-frequency image so that, effectively, a higher-frequency signal is synthesized with a lower-rate DAC. However, great attention still needs to be paid to the details associated with the many non-ideal effects discussed in Section 4.3, since the DAC circuitry's shortcomings have not been changed because of all this, and the output signal synthesized by the DAC is still subject to the impairments discussed previously.

4.7   Specialized DACs

In the previous sections current-steering DACs have been discussed in a general context. Some applications, however, put specific demands on the performance of the converter, and specialized current-steering DACs have been developed to meet such demands or optimize their performance accordingly. This section will give the reader a head start on some important aspects of recent specialized DACs in order to facilitate further independent delving into the specialized literature.

We will discuss specialized DACs for

- audio applications,
- direct digital synthesis (DDS),
- ultra-high-frequency (sometimes referred to as radio-frequency) applications such as in wireless and wireline application.

4.7.1   Advanced audio DACs

As discussed in Section 1.1.2, the output signals synthesized by audio DACs do not go far beyond the audible range of $\sim$20 kHz. Nevertheless, the required distortion and noise performance can be extremely demanding (e.g. THD + N > 100 dB in high-fidelity applications). That is often conjugated with restrictive specifications in terms of silicon area and power consumption since such DACs can be required to be integrated into complex systems-on-a-chip (SoCs) (hence in nanometer CMOS processes with low supply voltages) and employed in mobile applications [374].

Because of the above conditions, many such DACs are implemented using oversampled $\Delta\Sigma$ current-steering architectures since, compared with, for example, a traditional switched-capacitor architecture [375], the current-steering approach is less sensitive to on-chip coupling, results in lower area, is more amenable to lower supplies, and is a better fit to drive low impedance loads. On the other hand, the resulting noise floor is limited by the thermal noise introduced by the current sources of the DAC array, ISI introduces distortion (as well as further degrading the noise floor due to the introduction of several lower-power spurs that blend with actual noise), and sensitivity to the sampling clock jitter can be an additional challenge [354].

In order to take advantage of the intrinsic high speed offered by the process technology, oversampling is amply exploited with ratios of the order of $OSR = f_{out}/(2f_s) \geq 64$. Oversampled multi-bit DACs allow one to use a considerably lower number of DAC output levels than in a Nyquist-rate DAC, which also leads to much smaller current arrays than in the Nyquist-rate DACs discussed in previous sections. The accuracy of the output levels (mostly affected by static mismatch between the current sources) and their dynamic behavior (affected by ISI among other factors) is nevertheless very critical to achieving the desired THD; accuracy is controlled using dynamic element matching (DEM). To avoid undesired in-band down-mixing due to coupling and due to a jittery clock and minimize overall system cost and size, it is also important both to keep the spurious power outside the audible band (out-of-band spurs due both to aliases and to shaped quantization noise) to a low level and also to have a simple, low-power, I-to-V filtering[18] stage following the DAC. Thus the converter is preceded by suitable digital interpolators/filters ahead of the digital $\Delta\Sigma$ modulator and the DEM logic.

In addition to all that, the audio DAC described in [354] extends some of the above techniques by using a complementary three-level (+1, 0, −1) DAC cell instead of

---

[18] A linear and low-noise current-to-voltage relationship is required since the desired final output signal must be a voltage.

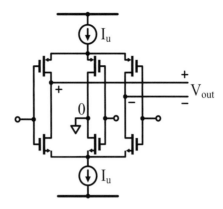

Figure 4.65 A three-level output DAC. Note that, when the zero level is converted, no actual current is sourced/connected to the output and the current sources are steered toward an internal AC ground [354]. © 2008 IEEE.

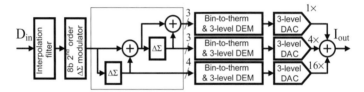

Figure 4.66 The block diagram for the audio DAC introduced in [354]. © 2008 IEEE.

a traditional two-level $(+1, -1)$ cell. That is conceptually exemplified in Fig. 4.65. The finer granularity introduced by the extra output level has a number of advantages, including

- lower thermal noise on the output since, when 0 is the output of a cell, no corresponding current is actually contributed to the load from the cell;
- having more levels overall, the individual current density of the cells can be reduced;
- synthesizing a low-power output results in less switching activity than in the two-level case and hence, overall, lower power consumption.

On the other hand, a three-level DEM is required, and a new algorithm extending to three levels a first-order noise-shaping DEM has been used [376]. A high-level block diagram of the DAC is depicted in Fig. 4.66.

Since the multi-bit DAC is still an 8 b DAC, the DEM is segmented by partitioning the 8 bits into smaller segments and then each segment has an individual DEM. However, to minimize the impact of errors introduced at the boundaries between the segments, these are further scrambled by means of first-order digital $\Delta\Sigma$ modulators. Finally, the DAC output is fed to an active I-to-V converter as shown in Fig. 4.67. V_{out} does not return to zero after the conversion phase. But the DAC's output current is made to change during part of the "hold" phase, hence masking associated nonlinear transients (eliminating the

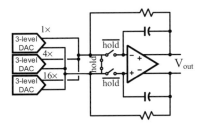

Figure 4.67 The output stage including the I-to-V low-noise conversion and DAC deglitching [354]. © 2008 IEEE.

corresponding distortion with this "deglitching" scheme) that happen during the switch state-change transition similarly to what has been described in the previous sections for RTZ DACs.

A further extension of this three-level DEM was introduced in [377], where, using a signed binary coding, the pointers for the DEM are updated in alternate directions, allowing more uniform usage of all elements and lower tonal behavior. Furthermore, greater power efficiency and lower noise are achieved through using a number of low-power techniques.

A new approach to DAC segmentation resembling a $\Delta\Sigma$ MASH was introduced in [378]. In this technique the MSB segment is first digitally noise-shaped; then the difference between the binary code driving the MSB segment and the original digital input (hence this difference is essentially the residue) is further digitally noise-shaped before using it to drive a suitably gain-scaled lower order segment. This can be further extended to lower and lower order segments by recursive noise-shaping and gain scaling of the corresponding digital residues.

Finally, most of the past DEM scramblers can noise-shape the distortion originating from sources of static nonlinearity (e.g. mismatch of DAC elements) but do not address sources of dynamic nonlinearity such as the previously discussed nonlinear ISI. The latter stems from introducing present output dependence from prior output values. In [379] a traditional noise-shaping DEM loop is extended by introducing an additional DEM loop that accounts for prior output code history and accordingly provides suitable dithering to the main DEM loop. In this way both the tonal behavior of the static DEM and the ISI distortion are dramatically reduced.

4.7.2 Direct digital synthesis

Direct digital synthesis (DDS) is another technique that makes use of high-speed current-steering DACs. Unlike arbitrary waveform generators (AWGs), wherein a traditional linear DAC such as those discussed previously is needed, in DDS the output of the DAC is always and only a sine wave. Various techniques to take advantage of this specific circumstance to optimize various performance dimensions (power consumption, area, linearity, and noise) have been reported [380].

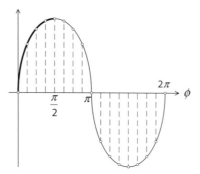

Figure 4.68 A sampled sine wave. The samples in the first quarter can be used as a basis for the generation of the entire waveform.

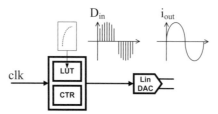

Figure 4.69 A DDS scheme based on a digital sequencer using a look-up table (LUT) with the base sine-wave samples, followed by a linear DAC.

A periodicity property of the sine wave, for example, has been widely exploited as explained in the following. A sampled sine wave is shown in Fig. 4.68. It can be noticed that the negative waveform in the second half of the period ($\phi \in [\pi, 2\pi]$) is simply a sign-reversed copy of the first half ($\phi \in [0, \pi]$). Furthermore, the second quarter of this wave ($\phi \in [\pi/2, \pi]$) is a mirror image of the first quarter ($\phi \in [0, \pi/2]$). Therefore, the absolute amplitude information on the sine wave is fully contained in the samples of the first quarter alone ($\phi \in [0, \pi/2]$). This observation is at the base of the "quarter-sine-wave compression" and has been exploited in various ways [381].

For instance, a DDS DAC can be built by using a standard linear DAC, such as those discussed in the previous sections, preceded by a digital sequencer using a ROM look-up table (LUT) with the samples of the first quarter sine wave and a suitable signed counter (CTR) scanning this table (and introducing the proper sign) to feed the DAC with the desired sample sequence as generically sketched in Fig. 4.69 [381, 382].

Although the principle is simple, its practical implementation can be plagued by several issues. To begin with, as a result of accuracy specifications, the size of the LUT can become very large due to two aspects:

- the resolution of the digital sine-wave samples in the LUT must be much higher than the DAC resolution in order for the sine-wave quantization to be limited by the DAC quantization error instead of the LUT quantization error;

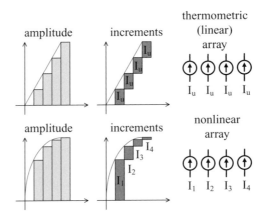

Figure 4.70 A DDS scheme based on nonlinear re-sizing of the current array to follow the increments of a sine wave. At the top a thermometric array is used to synthesize a linear ramp. At the bottom a sine wave is obtained by sequentially activating the nonlinearly sized currents of the second array.

- a low-frequency output sine wave at f_{out} will require several samples representing all the output levels when the output rate f_s is considerably larger (possibly, orders of magnitude larger) than f_{out}. As expected, the digital sequencer/accumulator scanning this LUT scales up accordingly.

In order to mitigate such issues and reduce both the digital word size and the number of samples stored in the LUT, several compression approaches have been proposed, including *CORDIC* [382, 383] and nonlinear interpolation [384, 385].

In [386] a $\Delta\Sigma$ digital interpolator is used to both compute and insert (by interpolation) extra samples in between those stored in the LUT (when required), as well as to increase their resolution by means of noise-shaping. The additional digital complexity required by the fourth-order phase-domain $\Delta\Sigma$ modulator, together with the fast signal processing enabled by clever digital design, is traded off for a significant reduction in the LUT size.

Another alternative is to "move the look-up table into the analog domain" by means of a post-D/A nonlinear phase-to-amplitude conversion [387, 388] or, more commonly, by means of a nonlinear DAC [389, 390, 391, 392]. In the latter case, just to fix the ideas, if a thermometric DAC, wherein each current source is nominally the same, is replaced by a DAC where the increments between each current source and the next one follow the incremental law of the steps of the quarter sine wave (see Fig. 4.70) then the sine can be synthesized by suitably sequencing the elements of this nonlinear array.

Although the latter approach, called ROM-less DDS, is, in principle, hardware-efficient and conducive to the implementation of very-fast-output-rate DACs, on the other hand, it is immediately apparent that, due to the large disparity in size among the increments, the required accuracy of the current sources results in very difficult matching specifications.

This problem can be attacked by segmentation techniques [390]. An alternate way to describe the segmentation principles explained in Section 4.1 for the case of linear

DACs is to say that higher accuracy is obtained through segmentation by using a finer DAC (e.g. an LSB segment) to recreate the intermediate steps between two large steps of the coarser DAC (e.g. the MSB segment). However, in the case discussed in Section 4.1, both DACs are linear in that they convert a linearly increasing digital input sequence into a correspondingly linear analog output waveform. The segmentation approach described in [393] for a ROM-less DDS DAC instead partitions the quarter sine wave into intervals where nonlinear finer DAC segments are used to fill in the gaps between larger-amplitude steps from the coarser DAC segments. This hence represents a nonlinear extension of the segmentation concept and allows one to mitigate the above-described accuracy/matching issue. This nonlinear approach relies heavily on trigonometric approximations and mimics in the analog domain similar decomposition approaches to those previously used in the digital domain to tackle the LUT size issue.

On the other hand, in [394], while the coarse segment uses a nonlinear DAC, the fine segment uses a linear DAC. This results in significant architectural simplifications in the converter and higher output rate, particularly considering that this is a 90 nm CMOS DAC (in contrast to some of the other cited implementations using BiCMOS and heterojunction processes).

Nonlinear DACs for DDS applications continue to be an active subject of research. Despite the intention of providing higher performance (in terms of area and power consumption, but also, more importantly, high-frequency linearity and noise) than a DDS system using a linear DAC, a significant net advantage over the latter has not yet been conclusively demonstrated. The two approaches continue to be in close competition, particularly with regard to dynamic performance.

Part of it can be explained on the basis of what has previously been discussed in Section 4.3. In fact, at high frequency, both approaches are similarly limited by common limitations such as timing skews, undesired coupling, and finite output impedance. Until a nonlinear DAC becomes significantly architecturally fitter to better tackle these high-frequency challenges, a clear superiority will not be possible.

4.7.3   RF DACs

A large part of the material discussed in this chapter has essentially been building up the foundations for one of today's most actively pursued areas of research centered around current-steering DACs, namely the so-called RF DACs. As the name suggests, this is a category of very-high-speed DACs for which the synthesized DAC's output signal is at radio frequency (RF), as opposed to being in baseband (BB) or at some intermediate frequency (IF) as in more established DAC applications.

The term RF is actually quite broad insofar as it classically encompasses all frequencies between, say, 30 kHz and 300 GHz. For example, traditional commercial FM radio stations broadcast in the range 87.5–108.0 MHz, which, by today's standards, isn't really a challenging output frequency range for a current-steering DAC in a nanometer CMOS technology. Indeed, the RF DACs we refer to in this section are those whose output signals are meant to generally reside beyond 1 GHz and for which it is intended to "move" the RF up-conversion stage of a transmitter into the digital domain. Such DACs

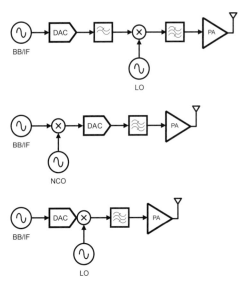

Figure 4.71 Different implementations of a transmit chain. Top, a traditional scheme with the DAC synthesizing the BB/IF signal, subsequently up-converted by an RF mixer and finally provided to the antenna by the power amplifier (PA). Center, a digital transmitter scheme whereby the BB/IF signal is digitally up-converted to RF using a digital mixer and a numerically controlled oscillator (NCO) before the RF DAC converts it into an analog input for the PA. Bottom, a mixed-signal merged DAC-plus-mixer scheme using a local oscillator (LO) for the RF carrier.

find application in wireless BTS applications as well as a variety of cable-based/wireline applications [14, 89, 90].

The challenges for Nyquist-rate current-steering DACs to meet the noise and distortion performance specifications for broadband (multi-channel/multi-carrier) RF output signals are formidable and have been covered extensively in the previous sections of this chapter, together with a variety of techniques with which to attack them. A sample of alternate approaches, some somewhat unorthodox, to this engineering problem is the topic of this section.

To begin with, one of the first papers explicitly introducing the RF DAC term was perhaps paper [396]. A simplified traditional up-conversion chain is sketched in the top scheme of Fig. 4.71.

One of the basic ideas in [396] consists of merging the function of the RF mixer with the DAC as generically depicted in the bottom scheme of Fig. 4.71. One observation that needs to be made immediately is the following. On comparing the traditional scheme at the top of Fig. 4.71 with the RF DAC scheme at the bottom of the same figure, it can be noticed that the reconstruction filter placed between the DAC and the mixer in the top scheme does not exist in the bottom scheme. This means that the images at the output of the DAC are not attenuated before being up-converted by the mixer, and therefore this additional spurious content and its additional mixed tones can be expected.

A direct circuit implementation of that can be seen in the simplified schematic representation shown in Fig. 4.72. In this scheme the tail current sources, which normally

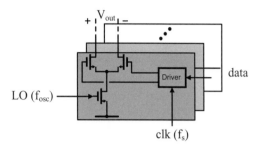

Figure 4.72 The principle of the mixed-signal RF DAC proposed in [396]. © 2004 IEEE.

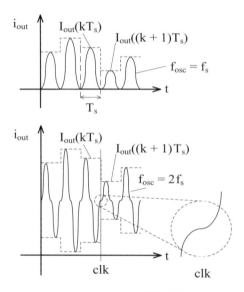

Figure 4.73 Output waveforms for the mixed-signal RF DAC for a sine pulse at $f_{osc} = f_s$ (top) and for an alternate sine pulse at $f_{osc} = 2f_s$ (bottom). Adapted from [396].

provide a constant current, are periodically modulated by the local oscillator (LO). Therefore the DAC's output current is indeed the result of mixing the converted digital input of the DAC with the LO's periodic waveform as shown in Fig. 4.73. For the waveforms represented in this figure two conditions have been imposed. First of all, the frequency of the LO f_{osc} has been chosen to be an integer multiple m of the output rate f_s, namely $f_{osc} = m \cdot f_s$. Moreover, the time at which the steering switches change state is set to be when the LO waveform has its peak, namely when its slope is zero and so any jitter on the DAC clock has minimal effect on the output waveform since the time uncertainty happens over a temporarily constant output. The latter relaxes considerably the aperture sampling clock sensitivity, which is particularly important at higher frequency [397]. It also greatly relaxes the distortion introduced by timing skews between the DAC elements. The above-mentioned relations between f_{osc} and f_s as well

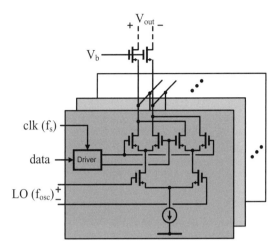

Figure 4.74 An alternate RF DAC cell [399]. © 2007 IEEE.

as the phase alignment between the LO waveform and the DAC clock can be obtained by means of a phase-lock loop.

The use of a pulsing (ideally a sine wave) waveform modulating the DAC's current sources has an additional advantage. While, as was seen in Section 4.6, a traditional rectangular/staircase output such as those in Figs. 4.50 and 4.51 leads to the sinc distortion (since the sinc function is the Fourier transform of the rectangular function), using a "smoother" output function like the sine wave leads to a more "benign" frequency distortion; one that does not attenuate the higher Nyquist images as much as the sinc function. In fact, it turns out, as shown in [396] that, similarly to what we previously discussed for the mixed-mode DAC output of Fig. 4.58, the alternating sine wave used for the DAC output shown at the bottom of Fig. 4.73 has a spectral distortion that suppresses the first Nyquist band image of the DAC input but enhances the second- and some of the higher-order images. That has the same benefits as the previously discussed mixed-mode DAC in that it allows one to use the filtered higher-order images as the intended output signal.

Finally, although the above principles can be applied to a regular Nyquist DAC, the implementation discussed in [396] refers to a $\Delta\Sigma$ DAC. This choice has been motivated by the intrinsic simplicity (significantly many fewer DAC cells) of a multi-bit $\Delta\Sigma$ DAC versus a Nyquist DAC, allowing a much better control of several sources of high-frequency performance loss (lower total parasitics, more compact layout, and hence better control of routing and timing issues, etc.). The main disadvantage of choosing a $\Delta\Sigma$ DAC is, however, the need to filter the out-of-band quantization noise. The latter makes the specifications of ultra-high-frequency bandpass filters following the DAC harder to meet than is the case with the traditional schemes.

A similar approach is described in [398, 399] (a Nyquist-rate architecture is proposed in [398], while a $\Delta\Sigma$ is discussed in [399]), where, however, the DAC cell is modified as shown in Fig. 4.74. It is interesting (and not surprising) to note how closely the DAC cells

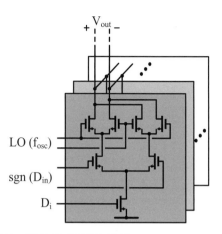

Figure 4.75 Digital RF DAC cell [402]. © 2009 IEEE.

of Figs. 4.72 and 4.74 actually resemble a single-balanced mixer and a double-balanced mixer, respectively [400]. This scheme helps with the rejection of the DC and even LO harmonic output, and the LO leakage is dominated by direct coupling. A complete transmitter IC utilizing this approach is described in [401].

Another variation of the latter topology is discussed in [402] and shown in Fig. 4.75. There are some differences worth mentioning. In this topology all transistors are actually used as switches (driven fully on or off) as in a digital circuit. The digital data is used to turn the tail current sources on and off, the polarity ($sgn(D_{in})$ in Fig. 4.75) drives the first level of switches, and, finally, the LO drives the top-level switching quads. The frequency ratio between f_{osc} and f_s as well as the phase alignment between the corresponding waveforms are governed by the same considerations as those pertaining to the first RF DAC scheme in order to minimize unwanted glitches and other associated impairments. Compared with the previous scheme, this provides a higher output power and better efficiency. But it also requires a larger LO driver and has higher LO leakage.

Since the current sources are turned on and off depending on the digital code D_{in}, the power consumption of the DAC cells scales with the magnitude of the digital input. At the same time, the biasing for the switches needs to stay turned on because it wouldn't be able to resettle at the operating switching frequency. As in all previous cases, care needs to be taken in the frequency planning and post-DAC filtering since the out-of-band spurious power can be significant.

The latter is one of the issues that the IC discussed in [403] addresses. This uses a *bandpass* $\Delta\Sigma$ modulator followed by an N-tap semi-digital FIR filter combined with 1-bit DACs to suppress undesired spurious content. Using 1-bit DACs avoids linearity issues associated with device mismatch and timing skew, while mismatches among them merely affect the transfer function and hence provide a relatively more robust architecture than using a multi-bit DAC [403]. Furthermore the linearity issues of conventional mixers are avoided in this approach, leaving the finite output impedance of the DAC cells as one of the most significant limitations.

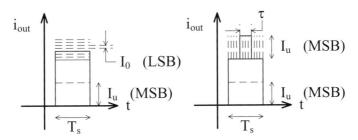

Figure 4.76 The output waveform obtained by adding up amplitude contributions from a current-mode MSB and an LSB segment (left-hand side) versus one obtained by adding up an amplitude contribution from a current-mode MSB segment and a pulse-width-modulated (time mode) LSB contribution (right-hand side) [404]. © 2010 IEEE.

Last but not least, a very different approach to increase the resolution of an RF-DAC is proposed in [404], where dynamic range is extended by applying a "digital-to-time" technique rather than a "digital-to-analog" conversion.

Without entering into the implementation details, the gist of this approach consists of increasing the granularity of an RF-DAC (or, more precisely, a digital-to-RF-amplitude converter (DRAC) as described in [404]) by implementing the LSB-segment conversion using pulse-width modulation (PWM) (i.e. the output stays on for a time that is proportional to the digital input) rather than the classic amplitude conversion of the LSBs (i.e. the amplitude of the output during the output sampling time is proportional to the digital input) as exemplified in Fig. 4.76. This approach is particularly attractive in the light of the fact that scaled nanometer CMOS processes challenge amplitude-based circuit techniques (due to the limited supply and voltage headroom) but offer the potential to exploit fine time resolution (thanks to the higher transition speed).

Finally, in applications for which a large dynamic range is not needed (SNDR ~ 25–30 dB or so), such as in wired backplane communication systems and optical communication systems, but ultra-high speed (tens of GSPS) is required, current-steering DACs composed of multiple parallel and identical sections each operating at lower rate but properly time-aligned to one another, similarly to time-interleaved ADCs, have been reported. With such techniques the proper alignment in time of the parallel sub-DACs is crucial, especially to prevent undesired glitches [395]. The other challenge consists of properly routing the data and control signals at several GSPS, particularly in CMOS [275, 303].

4.8   Conclusions

This chapter first provided a brief summary of established concepts of current-steering DACs, followed by some more recent results and new circuit approaches. Static and dynamic performance limitations of the basic architecture have been illustrated; circuit and layout techniques to address or mitigate them have been presented in the first sections of the chapter. Subsequent sections covered signal processing techniques increasingly being used in conjunction with this type of DAC, both to improve the converter performance and to enable specific system-level needs. Finally, some specialized DACs for important as well as emerging application areas have been discussed.

Although the technical literature in this area is somewhat fragmented and, perhaps, fewer people are actively involved with current-steering DACs than with some popular ADC architectures, many important technical challenges still severely limit the noise and distortion performance of this type of converter. The combination of a technological push toward finer CMOS processes and an application-based pull toward higher digitization of systems creates both the technical context and the need for DACs with ever higher-frequency output signals with ever greater signal quality. As a result of that, arguably, the last 5–10 years have probably seen more significant advances in this sense than did the previous decade. The work is, however, far from being even close to a state of maturity. To date, there are still many (deceptively) small implementation details, at circuit, layout, and packaging level, that can significantly impact the overall performance of a DAC that would otherwise give a respectable performance simply in terms of a first-order circuit-level simulator analysis.[19] Because of that, together with technical knowledge partly covered in this chapter, the designer's experience, thoroughness, and intuition are equally important ingredients for the success of a DAC design.

In the author's personal experience, learning to design and understand current-steering DACs has been somewhat similar to learning to speak English as a foreign language. Namely, it is relatively easy for many to speak and understand some basic English; but it becomes increasingly difficult to refine it and master it well enough to be eloquent and effective. Though the author can't claim to write elegant prose, it is the author's sincere hope that this chapter has taught the reader something new and interesting, opened the way to deepening knowledge further in the technical literature, and, more importantly, inspired researchers to further advancements in this exciting field.

---

[19] Simulation of these DACs is another tough and broad technical topic, plagued by challenges and hidden traps, that is beyond the scope of this book.

5   Trends in data conversion

It has been shown in Chapters 1 and 2 that the variety in data-converter specifications spans multiple decades of sample rate, distortion and noise performance, power consumption, area, etc. Different architectures are used to address subsets of this wide and diverse space. Some of the recent developments in architectures and design techniques have been covered in Chapters 3 and 4 and reflect the fact that, as technology evolves and commercial applications drive demands, the performance of converters has been varying over time. Observing and understanding the evolutions occurring within the context of a specific converter architecture, or within a particular application space, have also been part of what Chapters 3 and 4 have attempted to offer.

Do these "local" dynamics result in "global" trends? Are there common denominators that transcend individual architectures and result in aggregate observable evolutions? Is it possible to predict what the "state of the art" in data converters may be in years to come? Answering these and other, closely related, questions is extremely difficult and one of the lessons of the past is certainly that it is a very error-prone process. Nevertheless, some brave researchers, with the aid of historical data and performance models, have tried to spot trends and draw conclusions.

This chapter will summarize some of the recently published trends. An independent confirmation of some of these trends will be given. Moreover, additional and original predictions will be introduced and discussed.

5.1   Trends in ADCs

5.1.1   Performance trends

Notable surveys and analyses of ADC performance have been published by Walden [80], Murmann [82], and Jonsson [405]. In 1999, Walden surveyed many experimental and commercial ADCs and their main performance metrics (primarily SNR and SFDR versus sample rate). He introduced a now very popular figure of merit (the FOM_W of Eq. (2.10) in Chapter 2) and identified both some of the fundamental challenges (thermal noise, aperture uncertainty, comparator metastability) and some of the performance tendencies: roughly an improvement of 9 dB in SNR (and about 6 dB improvement in SFDR) between 1989 and 1997. Furthermore, this dynamic performance improvement

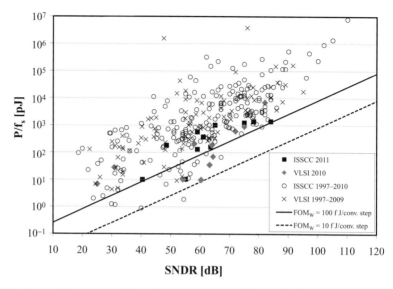

Figure 5.1 A recent "energy plot" from Murmann's survey [85]. © 2008 IEEE.

was not regular and it was, in fact, characterized by sporadic advances. This is to be expected since large changes in performance are often the result of breakthrough innovations in architecture, circuit design, or process technology. On the other hand, a considerably larger improvement in conversion energy efficiency (measured in terms of FOM_W) was visible over the same time span. He conjectured that much of that could be credited to [80]

- focus on monolithic ADC design, hence power efficiency was a primary goal;
- a recent and general de-emphasis on research and development;
- the existence of only a few application drivers pushing the state of the art (he highlighted software radio and satellite communication as key drivers for future breakthroughs).

About 9 years later, Murmann analyzed the performance data for all ADCs published at key technical conferences for this field (ISSCC and VLSI) between 1997 and 2008. Avoiding possible biases originating from specific figures of merit, the data was primarily presented and analyzed in the form of plots of "conversion energy" (P/f_s) versus SNDR (also known as "energy plots") and input signal bandwidth (BW) versus SNDR (also known as "aperture plots"). This survey is kept up to date and available online [85], and two recent such plots are shown in Figs. 5.1 and 5.2.

Murmann confirmed the visible overall progress in energy efficiency. Specifically, he quoted an average halving in power consumption every 2 years. Here we relay that information using the scatter plot of FOM_3 for the years 1997–2011 shown in Fig. 5.3.

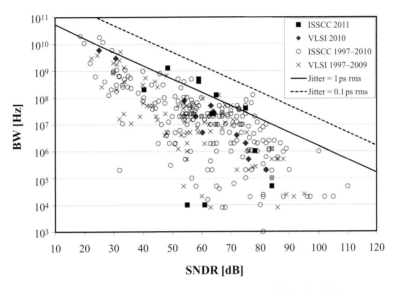

Figure 5.2 A recent "aperture plot" from Murmann's survey [85]. © 2008 IEEE.

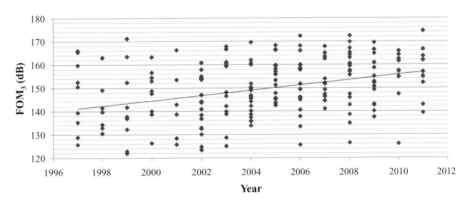

Figure 5.3 The conversion energy efficiency FOM_3 for the years 1997–2011 for all CMOS ADCs. A general doubling in energy efficiency every 2 years has been deduced.

Despite the efficiency improvement, Murmann highlighted too a rather negligible change in top-speed–resolution $(BW \cdot 2^{ENOB})$ performance during the past decade.[1] More on this $BW \cdot 2^{ENOB}$ "stagnation" will be discussed in the next section in conjunction with CMOS scaling. From the number of publications and what was reported in them

---

[1] The average trend indication of a doubling of $BW \cdot 2^{ENOB}$ every four years was actually accompanied by a large variance around this average, so the trend was judged to be only weakly conclusive. A comment was made in [51] that, in analogy to trends seen in other thriving industries, although pushing a specific technical performance parameter toward higher values would be technically possible, commercial motivations have driven performance evolution and research efforts in other directions.

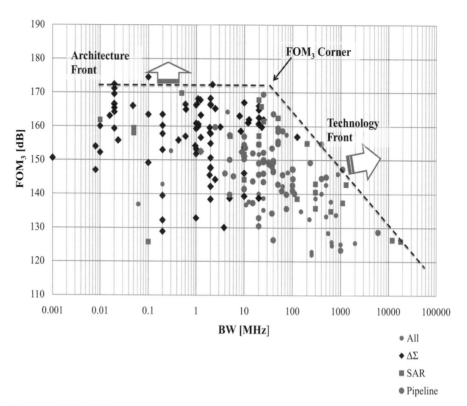

Figure 5.4 A plot of the conversion efficiency FOM$_3$ versus the input signal BW (courtesy of Hajime Shibata). Adapted From R. Schreier and G. Temes, *Understanding Delta–Sigma Data Converters*, IEEE Press/J. Wiley, 2005.

Murmann also noted that there was more visible research activity and performance progress in low-to-moderate-resolution ADCs, compared with a relatively more modest publication activity in the high-resolution space. Moreover, while plotting normalized energy $(P/f_s)/(P/f_s)_{min}$ versus SNDR, Murmann highlighted an "SNDR corner" close to SNDR $=75$ dB. Most recent converters at that time tended to "congregate" to the left of the corner (SNDR < 75 dB) and the trend was heading toward better energy efficiency (smaller $(P/f_s)/(P/f_s)_{min}$) with only minor activity to the right of the corner.

That is somewhat consistent with what had previously been shown by Schreier and Temes in [81] by plotting the conversion efficiency figure of merit FOM$_3$ (see Eq. (2.12) in Chapter 2) versus SNDR. A similar such plot created using the very same data set as that of Murmann's survey is shown in Fig. 5.4. An "FOM$_3$ corner" is visible in this plot for BW $\sim$ 30 MHz. Consistently with what was observed by Murmann in [82], lots of publication activity has been happening to the right of the FOM$_3$ corner (which tends to be characterized by wideband input signal bandwidth and medium-to-low-resolution ADCs) while those ADCs with narrower input bandwidth BW and higher SNDR gather around the horizonal dashed line. Over time, we have been witnessing a faster horizontal

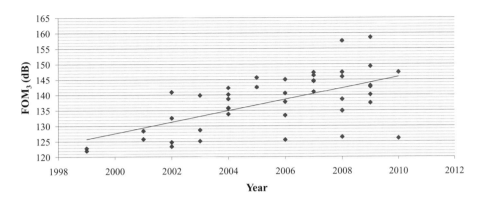

Figure 5.5 The conversion energy efficiency FOM_3 for the years 1997–2011 for high-speed ($2\,BW \geq 100\,MSPS$) CMOS ADCs only. An average $\sim$1.82 dB improvement in energy efficiency every year has been estimated.

shift toward the right of the diagonal dashed line and only a rather limited vertical shift up of the horizontal dashed line.

The horizontal dashed line was called the "architecture front" by Schreier and Temes [81]. The ADCs near it tend to have a high-SNR performance that is noise-limited, and it is their conversion-energy-efficient architecture rather than the process technology that determines any impact on the FOM_3. This front counts many $\Delta\Sigma$ ADCs and precision SAR ADCs, as shown in the figure. These types of ADCs are those generally referred to as "high-resolution" or "precision" ADCs.

Conversely, the diagonal dashed line was named the "technology front." Converters lining up along this border need to use less energy-efficient architectures (lower FOM_3) to meet their performance goals (very wideband operation and moderate SNR). These considerably faster converters, commonly referred to as "high-speed" ADCs, also rely on the speed capability of the devices, sometimes pushed to the process technology limits [406], in order to be able to hit their dynamic performance targets and, to achieve that, need to burn quite a bit more power. Pipelined ADCs, F&I ADCs, and flash ADCs are prominently represented along this front.

While still keeping in mind the previously mentioned broad data spread, a closer analysis of the evolution over time of conversion efficiency trends for the "high-speed" ADCs versus the "high-resolution" ADCs suggests [82]

- a doubling in energy efficiency $(P/f_s)/(P/f_s)_{min}$ every 1.6 years for high-speed ADCs (see also Fig. 5.5 for an FOM_3 plot over time);
- a doubling in energy efficiency $(P/f_s)/(P/f_s)_{min}$ every 5.4 years for high-resolution ADCs (see also Fig. 5.6 for an FOM_3 plot over time).

In truth, precision ADCs too have made significant progress, although it is somewhat less visible with the above methods of investigation since not many publications have reported such developments [164]. However, after a more careful observation of the

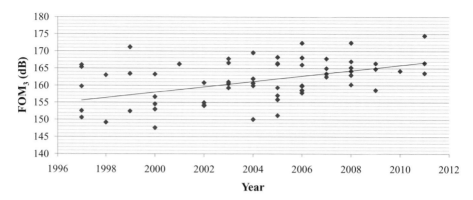

Figure 5.6 The conversion energy efficiency FOM$_3$ for the years 1997–2011 for high-resolution (SNDR $\geq$ 75 dB) CMOS ADCs only. An average $\sim$0.8 dB improvement in energy efficiency every year has been estimated.

specifications of commercial precision ADCs, meaningful information can be gathered. For instance, 16 b SAR ADCs have gone from sample rates of the order of 1 MHz in the early 2000s to almost ten times this speed in recent years. Combinations of the techniques described in Section 3.1.3, various types of calibration, trimming, and other proprietary techniques are behind that. Similar considerations apply for the very-high-precision ADCs, dominated by $\Delta\Sigma$ ADCs [407, 408]. Interestingly, converters in this resolution range were relatively common in academic publications during the early nineties.

Both in Walden's paper and in Murmann's paper, the contour lines of the aperture jitter equation (see also Eq. (2.17) in Chapter 2 for more discussion),

$$\text{SNR}_{\text{dB}} = 20 \log \left(\frac{1}{\sqrt{2} A \pi f_{\text{in}} \sigma_{\text{j}}} \right) \tag{5.1}$$

were highlighted in some of the performance plots.

Two of these contours are visible in the aperture plot in Fig. 5.2 and it should be remarked that, again in a time-discontinuous manner, some converters have moved through these boundaries over the years, going from the $\sigma_{\text{j}} \sim 1$ ps rms line in Walden's paper (1999) and coming very close to $\sigma_{\text{j}} \sim 0.1$ ps rms in the up-to-date plot shown in Fig. 5.2: almost a factor of 10 improvement in 12 years regarding a rather challenging and fundamental limitation that hinders converter architectural choices.

5.1.2   CMOS scaling

An important factor in determining trends is certainly CMOS process scaling. To begin with, a distinction needs to be made between stand-alone ADCs and embedded ADCs [3]. The former are primarily general-purpose ADCs for which a balanced SNR and THD performance is crucial and expected to be nearly constant through the entire input signal bandwidth BW. Conversely, in embedded ADCs, similar nominal SNDR performance, compared with that of general-purpose ADCs, can actually be found in ADCs that are very linear but highly noise-limited (THD $\gg$ SNR) or the other way around (SNR $\gg$ THD). Either of these two metrics may be optimized to be maximal only over certain signals or conditions [3], depending on the intended application. Lastly, thanks to the considerably larger availability of digital processing circuitry on such systems-on-a-chip (SoCs), it is not uncommon for embedded ADCs to make significant use of digital calibration and post-processing to compensate for the analog shortcomings typical of finer lithography processes.

Historically, for stand-alone ADCs the adopted process and process options have been those which provided the best trade-off among (analog) converter performance, costs, process/design infrastructure stability, and designer-related experience. More than 15 years ago state-of-the-art ADCs in this category typically adopted finer CMOS process geometries at a rate of halving the minimum channel length L every 5.4 years [405]. Such ADCs traditionally lagged state-of-the-art digital chips in process adoption by at least two or three process nodes.

During the same years, in order to be integrated into chips with predominant digital content, embedded ADCs were certainly closer, in terms of process adoption, to state-of-the-art purely digital CMOS nodes [3], albeit with a different type of performance. The rather scarce documentation available in the open literature on this subject doesn't allow accurate general assessments; however, in these cases, the adoption lag was more in the range of one or two process nodes, compared with, say, a cutting-edge DSP chip.

However, the adoption rate for published converters during the last 15 years has dramatically accelerated. During that time span, L halved every 3.7 years [405]. In fact, in very recent years, both stand-alone and embedded ADCs have rapidly been closing their process node gap and, certainly, their gap with mainstream digital chips.

The impact of CMOS process scaling on conversion efficiency has been discussed in the previous section and it has been remarked how less energy-efficient high-speed ADCs tend to benefit more from scaling than do their precision counterparts. An in-depth analysis on how different blocks and architectures are affected in different ways from scaling can be found in [28, 29, 31, 82, 406]. Scatter plots showing the relative FOM_3 improvements for all CMOS ADCs and then for high-speed ($2\,BW \geq 100\,MSPS$) and high-resolution (SNDR $\geq 75\,dB$) ADCs are reported in Figs. 5.7–5.9.

It has been remarked in the previous section that the $BW \cdot 2^{ENOB}$ product has not changed that much during the past 10–15 years or so. As pointed out before, part of the reason for that has certainly been application demand-related. However, another explanation is related to CMOS scaling. As supply voltages dropped over process nodes, due to headroom limitations, the allowed signal power had to decrease accordingly while the noise spectral density (NSD) has been kept nearly constant. Nodes with smaller feature sizes have allowed wider signal bands thanks to the increased processing speed. But, overall, the combination $BW \cdot 2^{ENOB}$ has seen the change in its different components essentially canceling each other out over process generations, leading to a largely unchanged $BW \cdot 2^{ENOB}$ product.

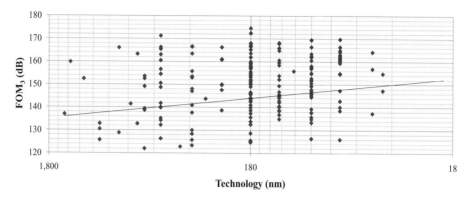

Figure 5.7 The conversion energy efficiency FOM_3 in CMOS process nodes from 1 μm to 40 nm for all CMOS ADCs. A general ~1.5 dB improvement in energy efficiency per technology node has been deduced.

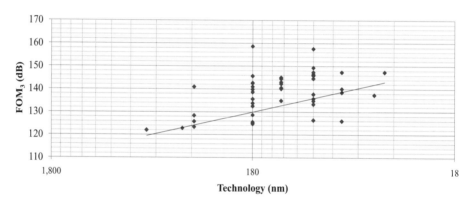

Figure 5.8 The conversion energy efficiency FOM_3 in CMOS process nodes from 1 μm to 40 nm for high-speed (2 BW ≥ 100 MSPS) CMOS ADCs only. An average ~6 dB improvement in energy efficiency per technology node has been deduced. Note the popularity of the 0.18 μm and 90 nm nodes. Also, several older designs have been ignored in this plot which is based on the somewhat arbitrary definition of high speed as 2 BW ≥ 100 MSPS.

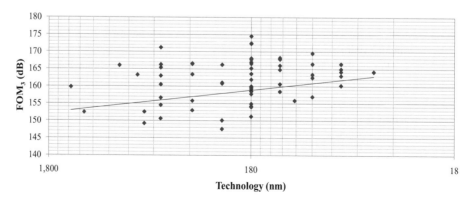

Figure 5.9 The conversion energy efficiency FOM_3 in CMOS process nodes from 1 μm to 40 nm for high-resolution (SNDR ≥ 75 dB) CMOS ADCs only. An average ~1.6 dB improvement in energy efficiency per technology node has been deduced.

5.1.3   Architectural and application considerations

A very pertinent observation has been made in [28], namely that CMOS scaling has led to a general trend toward lower-resolution designs for Nyquist converters. On the other hand, after reading some of the previous chapters, some other subtle trends can also be identified. For example, the increase in transition frequency f_T, the decrease in voltage headroom, and the increased availability of digital functionality, together with many of the other aspects discussed in Section 3.1.4, have created the conditions for a gradual architectural shift from Nyquist-rate ADCs to oversampled $\Delta\Sigma$ ADCs in the medium-resolution space between 10 and 14 ENOB in the next few years.

In [405] Jonsson reported a generalized and phased slowing down, over time, in the increase of the peak sampling rate for various classes of converters. For example, true 8 b converters (ENOB ~ 8) have shown a peak f_s increase going from fractions of MSPS in the mid eighties to a few hundred MSPS only 5 years later, but then saw hardly any further improvement in sample rate for the following 15 years in a row. A very similar trend, with a delay of about 5–7 years, although a bit slower and smoother, has also been reported for the peak sample rate of true 12 b converters and then also 14 b (and greater resolution) converters.

The reasons for these trends are partly due to CMOS scaling considerations similar to what has been witnessed on the digital side of the world: although integration continues to march in accord with Moore's law, the actual speed improvement offered by newer process generations is becoming gradually less aggressive. On the other hand, just as microprocessor and DSP designers are finding other avenues to leverage higher integration to obtain higher processing throughput, data converter designers aren't standing still either. Among the many innovative techniques and architectures discussed in Chapter 3, two of these, in particular, have recently been maturing to a level that can make real breakthroughs both in the bandwidth and in the frequency space for digitizing analog signals. These are time-interleaved ADCs and continuous-time $\Delta\Sigma$ ADCs. It is expectable that these two classes of converters, enabled by calibration and significant availability of digital functionality, will give a new acceleration to the rate at which BW has been growing.

It won't be at all surprising to see true 10 b converters sampling signals with BW $\sim$ 5 GHz or true 12 b converters sampling signals with BW $\sim$ 1 GHz or more within the next 5 years.

For such high frequencies and dynamic range, on the other hand, the challenge to tackle will be the sampling clock phase noise. This fundamental challenge hasn't to date seen real architectural breakthroughs allowing ADCs to get around the pace at which such pure clock sources become practically available and consistently priced with the converter costs.

Another factor that should be considered when analyzing the sample rate slow down of 8 b converters (followed by a similar pattern of a gradual sample rate slow down of 10 b converters 5–7 years later etc.) has little to do with device physics considerations and much to do with application considerations as well as how to interpret the raw data used in the analysis. For example, 8 b converters were mainstream during the eighties and nineties when their dramatic sample rate increase was witnessed and reported. However, over time, and certainly during the mid nineties, demand and availability of faster 10 b converters caused a shift from 8 b to 10 b in applications such as communication, instrumentation, data acquisition, medical applications etc. At the same time, 8 b ADCs have also begun to be more commonly used in a wide variety of embedded applications such as imaging and hard-disk drives, hence also becoming less commonly found in data-conversion-specific publications.[2] A similar shift is starting to happen between true 10 b and 12 b ADCs today, so it is expected that a substantial demand for true 12 b ADCs will drive a great deal of evolution for converters in this class in the next few years.

Much emphasis is, perhaps too often, placed on energy efficiency figures of merit in publications and among engineers in the data conversion community. A more holistic view of the many systems using ADCs and DACs, however, highlights that these components are neither the most power-hungry nor the least energy efficient among those sharing the same board. For example, in communication systems it is not uncommon that driving amplifiers placed in front of the ADC consume as much or even more power than the ADC itself, not to mention the power amplifier driving the antenna in a transmit path. This power amplifier dwarfs the DAC placed a few blocks behind it in terms of both power consumption and energy efficiency. Likewise, in a much broader application context, high-performance converters are often used in conjunction with series LDO supply regulators to insure cleaner supplies than in the considerably much more power-efficient switching regulators. If only the converter could better achieve high supply rejection (particularly at high frequency) then using a switching regulator would dramatically impact the overall system power consumption.

These two simple examples suggest that paying excessive attention to the energy efficiency of the converter alone could be a symptom of an inward-looking focus of the data converter design community [51]. If anything, this "FOM fever" could be more justified in rapidly expanding mobile applications where battery life is critical and in those applications where high power consumption leads to excessive die temperature and therefore loss of performance or, in extreme cases, prevents the integration of more circuitry on the same die or designing the converter itself in the first place. Packaging technology can also help with new low-thermal-resistivity θ_{JA} packages.

Finally, as data converter technology progresses, many new applications for ADCs emerge while others come into reach. For example, GSPS-rate low-resolution ADCs (around 6 b or less) are rapidly becoming commonplace in serial I/O receivers alongside more traditional DLLs/PLLs and other timing circuitry [409, 410]. Also, broad adoption of time-interleaving and other high-speed and -resolution architectures is bringing the "holy grail" of true software radio closer to reality.

Moreover, the borders between actual signal processing and ADCs might begin to blur. For example, data compression is being directly attached to the ADC, reducing bandwidth and storage requirements in applications like automated test equipment (ATE), medical imaging, and digital oscilloscopes [411]. Another example is digital filter banks and "blind calibration" being applied for calibration of time-interleaved ADCs [270, 412, 413].

---

[2] Which is the primary source of a lot of the data being analyzed for trend predictions.

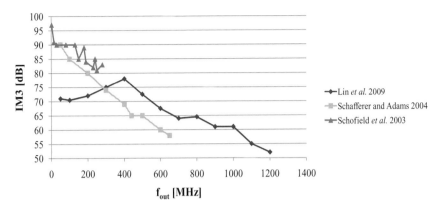

Figure 5.10 Third-order intermodulation distortion (IM_3) for three different DACs: Lin *et al.* 2009 [90] clocked at 2.9 GSPS, Schafferer and Adams 2004 [311] clocked at 1.4 GSPS, and Schofield *et al.* 2003 [298] clocked at 400 MSPS.

For those really in search of "new emotions," data converters have recently been implemented in flexible plastic and organic materials [414, 415, 416].

5.2 Trends in DACs

Assessing the performance of DACs is a much more complex matter than in the case of ADCs. While in most stand-alone ADCs the dynamic performance stays approximately constant all the way to the Nyquist frequency, the same is certainly not true for DACs.

Let us consider, for example, the third-order intermodulation distortion (IM_3) for three different current-steering DACs with relatively similar resolution and sample rate, shown in Fig. 5.10.

As discussed in Section 4.3, the dynamic linearity of current-steering DACs degrades very rapidly with increasing frequency as a result of the combination of various non-idealities.[3] However, due to the different ways in which such factors combine, each DAC has a slightly different frequency behavior as evidenced by the three different shapes of the plots in Fig. 5.10.

To make things worse, the conditions under which such performance is measured and reported vary quite a bit and make it very difficult to perform objective comparisons.

For instance, in the case of the three DACs of Fig. 5.10, the maximum output voltage swing is sufficiently different to have a meaningful impact on the linearity: 2.5 V_{pp} for [90], 1.5 V_{pp} for [311], and 1 V_{pp} for [298], respectively. Although, in principle, in the case of IM_3 a normalization could be applied by considering that IM_3 should improve

---

[3] In this specific example, assuming that non-idealities such as transistor mismatch, $r \cdot i$ drops, undesired coupling, timing skews, etc. have all been minimized by careful design and layout, it is reasonable to guess that the IM_3 performance of the DACs corresponding to the plot with diamond markers [90] and the one with square markers [311] will be limited by finite output impedance as discussed in Section 4.3.1 and as evidenced by the high-frequency degradation of IM_3 and its slope. The very same overall trend is also visible for the DAC with the plot with triangle markers [298], although also something else might be happening since it wiggles up and down quite a bit.

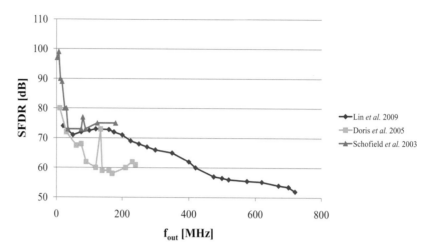

Figure 5.11 Spurious free dynamic range (SFDR) for three different DACs: Lin *et al.* 2009 [90] clocked at 1.6 GSPS, Doris *et al.* 2005 [334] clocked at 500 MSPS, and Schofield *et al.* 2003 [298] clocked at 400 MSPS.

by 3 dB for a decrease by 1 dB of the power of both of the two tones [78], it is sometimes hard to know the conditions under which the reported IM_3 has been measured. (For example, what is the power of the two tones corresponding to the reported IM_3 plots? This is not the case for the three DACs used in this example, but it is all too often the case in many other instances.) Moreover, it is also often unclear how far the power scaling ratio 3:1 applies and when it is in fact more like 1:1 (see Section 2.2.2).

The example of IM_3 is actually relatively simple. Matters become even more complex when trying to compare SFDR performance. An example is shown in Fig. 5.11, again with three DACs with similar resolution and output rate. Here the three SFDR trends are even more different from one another. Besides the fact that, again, the full scales are different, the SFDR is a very difficult metric to unravel when dealing with current-steering DACs because it is never clear, without seeing the corresponding FFT, which spurious content is determining it at specific frequencies. It could be dominated by third-harmonic distortion for low frequency, by fifth in an intermediate range, and by third again or by some other inharmonic tone at higher frequency. So, granted that it will surely degrade with increasingly higher frequency, the SFDR can vary somewhat unpredictably over narrow frequency ranges.

It is because of the above that, for example, an accepted figure of merit for DACs has not seriously emerged, and those found in the literature (see also Section 2.3) hardly capture the overall performance with a single metric. In practice, comparisons need to be made by superimposing entire performance curves as in Figs. 5.10 and 5.11 and hoping that the measurement conditions have been clarified sufficiently.

Granted all these considerations, it is interesting to observe the increase, over time, of the maximum output frequency at which a DAC could still perform, for example, to a 70 dB SFDR (we will call it f_{70}). That is shown in Fig. 5.12 and it corresponds to an increase of f_{70} by a factor of about 3.2 every 2 years. This suggests the possibility of a

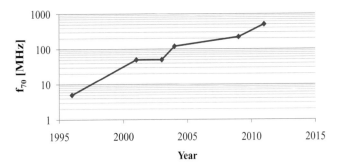

Figure 5.12 The highest frequency at which top communication DACs can still perform to 70 dB SFDR, over the course of the last 15 years. Data points are for [325] in 1996, [329] in 2001, [298] in 2003, [311] in 2004, [90] in 2009, and [321] in 2011.

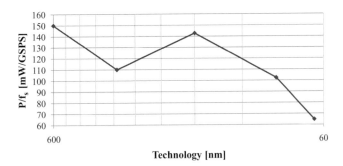

Figure 5.13 Representative data of energy efficiency (P/f_s) over subsequent generations of CMOS processes. Data points are from [325] in 0.6 μm, [329] in 0.35 μm, [311] in 0.18 μm, [321] in 90 nm, and [90] in 65 nm.

DAC performing at this level of linearity up to about 5 GHz (f_{out}, not f_s!) by 2016 or so. It should be remarked that, particularly in the most recent cases reported here, the improvement in linearity is primarily accomplished through circuit innovation and only in minor part contributed by the particular devices and process technology used. In fact, for example, the last case [321] benefits significantly from calibration and uses a 90 nm CMOS process.

This plot allows us to quantify a clear trend, namely the push to sustain high linearity at the highest possible frequency. The latter is crucial, for example, in wired and wireless communication applications as the digital/analog interface is moved closer and closer to the final power amplifier.

Energy efficiency trends are clearly just as interesting as those previously seen for ADCs. The plot of Fig. 5.13 reports some representative examples of the improvement of energy efficiency (P/f_s) over multiple CMOS process nodes. Despite the relatively reduced number of points, the overall trend attests to the improvement of the dynamic power consumption, which primarily comes from the digital circuitry in the encoding/ decoding logic but also in the switch control circuitry and internal signal routing. The

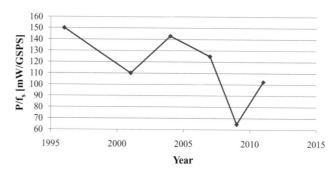

Figure 5.14 Representative data of energy efficiency (P/f_s) over the course of the last 15 years. Data points are from [325] in 1996, [329] in 2001, [311] in 2004, [89] in 2007, [90] in 2009, and [321] in 2011.

power associated with the static current sources is becoming a smaller contributor, particularly since, for the cases here surveyed, the full-scale current is right around the typical 20 mA, and (unlike the digital processing blocks) the stack of current sources and switches is powered with legacy supplies of the order of 3.3 V or so.

The plot of Fig. 5.14 reports similar energy efficiency over the course of the last 15 years. Again, despite its jagged shape, the overall trend confirms that there has been a gradual improvement in energy efficiency (P/f_s) over time.

On combining the trends shown in these two pictures, it is possible to project an energy efficiency of ~20 mW/GSPS in a 22 nm CMOS process in 2018.

Another trend that is primarily observable in commercial products rather than in academic publications is the increasing level of integration resulting from placing high-performance DAC cores together with digital signal pre-processing on the same die. During the last 5 years or so it has become customary for all the major producers of these devices to first release their state-of-the art DAC as a stand-alone core for general-purpose use, and soon to follow up with single and dual (possibly more) versions integrating digital interpolation filters, sinc pre-emphasis, digital up-conversion mixers etc.

As discussed in Sections 4.6 and 4.7, this is justified by the need for higher integration as well as ease of use. But it is also beneficial to the dynamic performance of the DACs because it moves the noisy high-speed data busses (e.g. the 14 b to 16 b data input busses D_{in} at full output rate f_s) from the chip's I/O pins into on-chip nodes with capacitive loads orders of magnitude smaller. The data rate at the chip's input of these highly integrated digital front ends is much lower thanks to the multiple on-chip interpolations, and having serialized input busses is also becoming a desirable feature to keep package pin count and board-level routing under control.

Similarly to what has been discussed for ADCs, recently there have been only very rare cases of academic publications in the area of precision D/As. Careful observation of trends in commercial parts, however, makes it possible to note typical top accuracy levels of about 16 b in the early nineties and through the early 2000s, followed by a significant improvement to around 20 b in very recent products.

6 Conclusions

As stated elsewhere in this book, a number of application drivers could potentially trigger important advances in the data conversion landscape in the short term. For instance, in the consumer market, the booming popularity and increasing pervasiveness of personal mobile devices (e.g. smartphones, tablet computers etc.) creates an unprecedented demand for data throughput in the wireless market that is somewhat similar to what has previously happened in the wired world when the internet became widely accessible. The same types of devices also drive a multitude of needs in terms of sensory interfaces and associated converters. Still in the consumer market, the gaming industry is also becoming both more technically sophisticated and more pervasive (which perhaps begs some societal questions) and leading to a considerably higher degree of interactivity and refined human–machine sensory interfaces, again relying on the ability to leverage fast DSP capability and, correspondingly, specialized ADCs and DACs.

TVs have finally made the technological shift from analog to digital, now also with high definition and even 3D. That perhaps means fewer interfaces with legacy analog systems, but more demand for data throughput over cable and fiber optics as well as, wirelessly, in home/apartment localized networks delivering all this digital content to various types of client devices and without passing cables through walls.

Automotive is yet another application area that is rapidly accelerating the demand for connected electronics and sensory networks. Various control and monitoring systems (e.g. pressure, temperature, speed), safety and driving aids (e.g. pre-crash detection, parking systems), internal and external connectivity, entertainment etc. are rapidly becoming standard features of modern cars and leading ultimately to demand for data converters.

Inevitably, in such an increasingly connected world, there will be positive and peaceful evolution but also unrest and conflicts. Sadly, the defence industry has been increasingly demanding larger bandwidth and deeper dynamic range for a multitude of "defensive" applications in harsh environments.

All these and many further examples of applications continue to fuel the seemingly unstoppable progression of Moore's law. Ironically, despite the relentless march toward finer CMOS lithography and growing amounts of digital processing functionality, analog and mixed-signal circuits have been taking on ever more diverse and, sometimes, unexpected embodiments, rather than "succumbing" to this digital supremacy. Very old ideas, which perhaps weren't technologically ready for fully successful implementation, together with rejuvenated architectures and entirely new types of circuit, have recently been emerging and perhaps replacing more established ones that have begun to struggle in the face of the restrictions of deep nanometer-scale technologies.

Among these, data converters continue to be the analog of microprocessors: just like their mighty digital counterparts, ADCs and DACs represent the most complex, sophisticated, and comprehensive analog and mixed-signal circuits. A modern high-performance data converter has got it all: amplifiers, comparators, references, filters, ... all the family members of analog design in a single system. That's fertile ground for competent and innovative generations of IC design engineers.

In conclusion, data converters surround us and their development will continue to be full of technical challenges and application drives for a very long time. It is the author's intention and hope that this book, despite the deserving topics for which space couldn't be found and those which could have been discussed in greater depth, has both helped the reader to gain a broader insight into today's (and maybe tomorrow's) converter landscape and stimulated further intellectual curiosity and engineering ideas that may lead to further advances and innovation.

References

[1] S. Inouye, M. Robles-Bruce, and M. Scherer, "2009 data converters," Databeans Inc., Reno, NV, technical report, June 2009.

[2] S. Inouye, M. Robles-Bruce, and M. Scherer, "2010 data converters," Databeans Inc., Reno, NV, technical report, June 2010.

[3] K. Bult, "Embedded analog-to-digital converters," in *European Solid-State Circuits Conference*, 2009, pp. 52–64.

[4] W. Kester, "Precision measurement and sensor conditioning," in *The Data Conversion Handbook*, W. Kester, Ed. Elsevier, Burlington, MA, 2005, pp. 539–561.

[5] R. L. Geiger, P. E. Allen, and N. R. Strader, *VLSI: Design Techniques for Analog and Digital Circuits*, Electronic Engineering Series. McGraw-Hill, New York, NY, 1990.

[6] J. Márkus, J. Silva, and G. C. Temes, "Theory and applications of incremental Δ–Σ converters," *IEEE Transactions on Circuits and Systems – Part I*, vol. 51, no. 4, pp. 678–690, 2004.

[7] W. Kester, "Digital audio," in *The Data Conversion Handbook*, W. Kester, Ed. Elsevier, Burlington, MA, 2005, pp. 591–605.

[8] J. P. Conti, "Wake up to wireless," *IET Engineering & Technology*, vol. 5, no. 13, pp. 14–15, 2010.

[9] A. Sheikholeslami, "ISSCC 2011 forum on high-speed transceivers: Standards, challenges, and future," in *International Solid-State Circuits Conference*, IEEE, 2011.

[10] W. Kester, "Software radio and IF sampling," in *The Data Conversion Handbook*, W. Kester, Ed. Elsevier, Burlington, MA, 2005, pp. 633–666.

[11] V. K. Garg and J. E. Wilkes, *Wireless and Personal Communication Systems*. Prentice Hall, New York, NY, 1996.

[12] P. Hendriks, "Specifying communications DAC's," *IEEE Spectrum*, vol. 34, no. 7, pp. 58–69, 1997.

[13] C. Edwards, "In the balance," *IET Engineering & Technology*, vol. 5, no. 14, pp. 60–62, 2010.

[14] G. Manganaro and D. Leenaerts, "ISSCC 2011 forum on advanced transmitters for wireless infrastructure," in *International Solid-State Circuits Conference*, IEEE, 2011.

[15] A. A. Abidi, "The path to the software-defined radio receiver," *IEEE International Journal of Solid State Circuits*, vol. 42, no. 5, pp. 954–966, 2007.

[16] P. B. Kenington and L. Astier, "Power consumption of A/D converters for software radio applications," *IEEE Transactions on Vehicular Technology*, vol. 49, no. 2, pp. 643–650, 2000.

[17] J. Sevenhans and Z.-Y. Chang, "A/D and D/A conversion for telecommunication," *IEEE Circuits & Devices Magazine*, vol. 14, no. 1, pp. 32–42, 1998.

[18] L. Breems and J. H. Huijsing, *Continuous-Time Sigma–Delta Modulation for A/D Conversion in Radio Receivers*. Kluwer, Dordrecht, the Netherlands, 2001.

[19] EE Times editors, "Medical electronics gets personal," http://www.nxtbook.com/nxtbooks/cmp/eetimes_medelec_20101108/\#/3/OnePage, 2010.

[20] C. Hammerschmidt, "National Semi to commercialize sigma delta ADC," http://www.eetimes.com/electronics-news/4190836/National-Semi-to-commercialize-Sigma-Delta-ADC, 2008.

[21] C. Stagni, C. Guiducci, L. Benini *et al.*, "CMOS DNA sensor array with integrated A/D conversion based on label-free capacitance measurement," *IEEE International Journal of Solid State Circuits*, vol. 31, no. 12, pp. 2956–2964, 2006.

[22] A. C.-W. Wong, D. McDonagh, G. Kathiresan *et al.*, "A 1 V, micropower system-on-chip for vital-sign monitoring in wireless body sensor networks," in *International Solid-State Circuits Conference*, 2008, pp. 138–139.

[23] S. Chou, "Integration and innovation in the nanoelectronics era," in *International Solid-State Circuits Conference*, 2005, pp. 36–41.

[24] T.-C. Chen, "Where CMOS is going: Trendy hype vs. real technology," in *International Solid-State Circuits Conference*, 2006, pp. 22–28.

[25] G. Moore, "Cramming more components onto integrated circuits," *Electronics*, vol. 38, no. 8, pp. 114–117, 1965.

[26] G. Moore, "No exponential is forever: But 'forever' can be delayed!," in *International Solid-State Circuits Conference*, 2003, pp. 20–23.

[27] C. Edwards, "Analogue circuits squeeze chip design," *IET Engineering & Technology*, vol. 5, no. 8, pp. 14–15, 2010.

[28] Y. Chiu, B. Nikolić, and P. R. Gray, "Scaling of analog-to-digital converters into ultra-deep-submicron CMOS," in *Custom Integrated Circuits Conference*, 2005, pp. 375–382.

[29] K. Bult, "The effect of technology scaling on power dissipation in analog circuits," in *Analog Circuit Design*, M. Steyaert, A. H. van Roermund, and J. H. Huijsing, Eds. Springer-Verlag, Berlin, Germany, 2006, pp. 251–290.

[30] K. Bult, "Analog design in deep sub-micron CMOS," in *European Solid-State Circuits Conference*, 2000, pp. 126–132.

[31] B. Murmann, P. Nikaeen, D. J. Connelly, and R. W. Dutton, "Impact of scaling on analog performance and associated modeling needs," *IEEE Transactions on Electron Devices*, vol. 53, no. 9, pp. 2160–2167, 2006.

[32] M. Boulemnakher, E. Andre, J. Roux, and F. Paillardet, "A 1.2 V 4.5 mW 10 b 100 MS/s pipeline ADC in a 65 nm CMOS," in *International Solid-State Circuits Conference*, 2008, pp. 250–251.

[33] D. Johns and K. Martin, *Analog Integrated Circuit Design*. John Wiley & Sons, New York, NY, 1997.

[34] F. Maloberti, *Data Converters*. Springer-Verlag, Berlin, Germany, 2009.

[35] A. Thomsen, D. Kasha, and W. Lee, "A five stage chopper stabilized instrumentation amplifier using feedforward compensation," in *VLSI Circuits Conference*, 1998, pp. 220–223.

[36] J. Yan and R. L. Geiger, "Fast-settling amplifier design using feedforward compensation technique," in *IEEE Midwest Symposium on Circuits and Systems*, 2000, pp. 494–498.

[37] M. Moyal, M. Groepl, H. Werker, G. Mitteregger, and J. Schambacher, "A 700/900 mW/channel CMOS dual analog front-end IC for VDSL with integrated 11.5/14.5 dBm line drivers," in *International Solid-State Circuits Conference*, 2003, pp. 416–417.

[38] J. Harrison and N. Weste, "A 500 MHz CMOS anti-aliasing filter using feed-forward op-amps with local common-mode feedback," in *International Solid-State Circuits Conference*, 2003, pp. 132–133.

[39] B. K. Thandri and J. Silva-Martínez, "A robust feedforward compensation scheme for multistage operational transconductance amplifiers with no Miller capacitors," *IEEE International Journal of Solid State Circuits*, vol. 38, no. 2, pp. 237–243, 2003.

[40] R. Ito and T. Itakura, "Phase compensation techniques for low-power operational amplifiers," *IEICE Transactions on Electronics*, vol. 93-C, no. 6, pp. 730–740, 2010.

[41] A. M. Abo and P. R. Gray, "A 1.5-V, 10-bit, 14.3-MS/s CMOS pipeline analog-to-digital converter," *IEEE International Journal of Solid State Circuits*, vol. 34, no. 5, pp. 599–606, 1999.

[42] M. J. Pelgrom, A. C. J. Duinmaijer, and A. P. G. Welbers, "Matching properties of MOS transistors," *IEEE International Journal of Solid State Circuits*, vol. 24, no. 5, pp. 1433–1440, 1989.

[43] P. R. Kinget, "Device mismatch and tradeoffs in the design of analog circuits," *IEEE International Journal of Solid State Circuits*, vol. 40, no. 6, pp. 1212–1224, 2005.

[44] P. G. Drennan and C. C. McAndrew, "Understanding MOSFET mismatch for analog design," *IEEE International Journal of Solid State Circuits*, vol. 38, no. 3, pp. 450–456, 2003.

[45] B. P. Wong, F. Zach, V. Moroz *et al.*, Eds., *Nano-CMOS Design for Manufacturability*. John Wiley & Sons, New York, NJ, 2009.

[46] L.-T. Pang and B. Nikolić, "Measurements and analysis of process variability in 90 nm CMOS," *IEEE International Journal of Solid State Circuits*, vol. 44, no. 5, pp. 1655–1663, 2009.

[47] J. Hurwitz, "ISSCC 2011 tutorial on layout: The other half of nanometer analog design," in *International Solid-State Circuits Conference*, IEEE, 2011.

[48] A. M. A. Ali, A. Morgan, C. Dillon *et al.*, "A 16 b 250 MS/s IF-sampling pipelined A/D converter with background calibration," *IEEE International Journal of Solid State Circuits*, vol. 45, no. 12, pp. 2602–2612, 2010.

[49] R. Payne, M. Corsi, D. Smith, S. Kaylor, and D. Hsieh, "A 16 b 100-to-160 MS/s SiGe BiCMOS pipelined ADC with 100 dBFS SFDR," in *International Solid-State Circuits Conference*, 2010, pp. 294–295.

[50] D. Robertson and T. Montalvo, "Issues and trends in RF and mixed signal integration and partitioning," *IEEE Communications Magazine*, vol. 46, no. 9, pp. 52–56, 2008.

[51] D. Robertson and M. Kessler, "Smart partitioning," in *The Data Conversion Handbook*, W. Kester, Ed. Elsevier, Burlington, MA, 2005, pp. 273–280.

[52] B. Razavi, *Principles of Data Conversion System Design*. IEEE Press, Piscataway, NJ, 1995.

[53] R. J. van de Plassche, *CMOS Integrated Analog-to-Digital and Digital-to-Analog Converters*, 2nd edn. Kluwer, Dordrecht, the Netherlands, 2003.

[54] B. Murmann, *EE315: VLSI Data Conversion Circuits – Class Notes*. Stanford University, Stanford, CA, 2008.

[55] W. Kester, "The importance of data converter static specifications – don't lose sight of the basics!," Analog Devices, Tutorial MT-010, 2009.

[56] Maxim Staff, "INL/DNL measurements for high-speed analog-to-digital converters (ADCs)," Maxim Integrated Products, Application Note 283, 2001.

[57] P. M. Figueiredo and J. C. Vital, *Offset reduction techniques in high-speed analog-to-digital converters*. Springer-Verlag, Berlin, Germany, 2009.

[58] B. Brannon, "Overcoming converter nonlinearities with dither," Analog Devices, Application Note AN-410, 1995.

[59] W. Kester and J. Bryant, "Data converter AC errors," in *The Data Conversion Handbook*, W. Kester, Ed. Elsevier, Burlington, MA, 2005, pp. 83–122.

[60] W. Kester, Ed., *The Data Conversion Handbook*. Elsevier, Burlington, MA, 2005.

[61] L. A. Singer and T. L. Brooks, "A 14-bit 10-MHz calibration-free CMOS pipelined A/D converter," in *VLSI Circuits Conference*, 1996, pp. 94–95.

[62] J. Li, R. LeBoeuf, M. Courcy, and G. Manganaro, "A 1.8V 10b 210MS/s CMOS pipelined ADC featuring 86dB SFDR without calibration," in *Custom Integrated Circuits Conference*, 2007, pp. 317–320.

[63] D. A. Mercer, "Low-power approaches to high-speed current-steering digital-to-analog converters in 0.18-μm CMOS," *IEEE International Journal of Solid State Circuits*, vol. 42, no. 8, pp. 1688–1698, 2007.

[64] A. Hastings, *The Art of Analog Layout*, 2nd edn. Prentice Hall, New York, NY, 2005.

[65] C. C. Enz and G. C. Temes, "Circuit techniques for reducing the effects of op-amp imperfections: Autozeroing, correlated double sampling, and chopper stabilization," *Proceedings of the IEEE*, vol. 84, no. 11, pp. 1584–1614, 1996.

[66] J. Yang and H.-S. Lee, "A CMOS 12-bit 4 MHz pipelined A/D converter with commutative feedback capacitor," in *Custom Integrated Circuits Conference*, 1996, pp. 427–430.

[67] I. Galton and P. Carbone, "A rigorous error analysis of D/A conversion with dynamic element matching," *IEEE Transactions on Circuits and Systems – Part II*, vol. 42, no. 12, pp. 763–772, 1995.

[68] I. Galton, "Why dynamic-element-matching DACs work," *IEEE Transactions on Circuits and Systems – Part II*, vol. 57, no. 2, pp. 69–74, 2010.

[69] S. Devarajan, L. Singer, D. Kelly *et al.*, "A 16-bit, 125 MS/s, 385 mW, 78.7 dB SNR CMOS pipeline ADC," *IEEE International Journal of Solid State Circuits*, vol. 44, no. 12, pp. 3305–3313, 2009.

[70] B. Grob and C. Herndon, *Basic Television and Video Systems*, 6th edn. McGraw-Hill, New York, NY, 1998.

[71] I. A. Chaudhry, S.-U. Kwak, G. Manganaro, M. Sarraj, and T. L. Viswanathan, "A triple 8 b, 80 MSPS, 3.3 V graphics digitizer," in *IEEE International Symposium on Circuits and Systems*, vol. 5, 2000, pp. 557–560.

[72] J. Li, G. Manganaro, M. Courcy, B.-M. Min, L. Tomasi, A. Alam, and R. Taylor, "A 10b 170 MS/s CMOS pipelined ADC featuring 84dB SFDR without calibration," in *VLSI Circuits Conference*, 2006, pp. 226–227.

[73] R. C. Taft, C. A. Menkus, M. R. Tursi, O. Hidri, and V. Pons, "A 1.8-V 1.6-GSample/s 8-b self-calibrating folding ADC with 7.26 ENOB at Nyquist frequency," *IEEE International Journal of Solid State Circuits*, vol. 39, no. 12, pp. 2107–2115, 2004.

[74] M. Gustavsson, J. J. Wikner, and N. N. Tan, *CMOS Data Converters for Communications*. Kluwer, Dordrecht, the Netherlands, 2000.

[75] B. Metzler, *Audio Measurement Handbook*, 1st edn. Audio Precision, Inc., Beaverton, OR, 1993.

[76] C. Motschenbacher and J. Connelly, *Low Noise Electronic System Design*. John Wiley & Sons, New York, NY, 1993.

[77] J. Munson, "Understanding high-speed DAC testing and evaluation," Analog Devices, Application Note AN-928, 2008.

[78] P. Wambacq and W. Sansen, *Distortion Analysis of Analog Integrated Circuits*. Kluwer, Dordrecht, the Netherlands, 1998.

[79] B. Brannon, "Sampled systems and the effects of clock phase noise and jitter," Analog Devices, Application Note AN-756, 2004.

[80] R. H. Walden, "Analog-to-digital converter survey and analysis," *IEEE Journal on Selected Areas in Communications*, vol. 17, no. 4, pp. 539–550, 1999.

[81] R. Schreier and G. C. Temes, *Understanding Delta–Sigma Data Converters*. John Wiley & Sons, New York, NY, 2005.

[82] B. Murmann, "A/D converter trends: Power dissipation, scaling and digitally assisted architectures," in *Custom Integrated Circuits Conference*, 2008, pp. 105–112.

[83] H.-S. Lee and C. G. Sodini, "Analog-to-digital converters digitizing the analog world," *Proceedings of the IEEE*, vol. 96, no. 2, pp. 323–334, 2008.

[84] B. E. Jonsson, "A survey of A/D-converter performance evolution," in *IEEE International Conference on Electronics, Circuits, and Systems*, 2010, pp. 768–771.

[85] B. Murmann, "ADC performance survey 1997–2011," http://www.stanford.edu/~murmann/adcsurvey.html, 2011.

[86] P. Palmers and M. S. J. Steyaert, "A 10-bit, 1.6 GS/s 27-mW current-steering D/A converter with 550-MHz 54-dB SFDR bandwidth in 130-nm CMOS," *IEEE Transactions on Circuits and Systems – Part I*, no. 11, pp. 2870–2879, 2010.

[87] A. Van den Bosch, M. S. J. Steyaert, and W. Sansen, "Solving static and dynamic performance limitations for high-speed D/A converters," in *Analog Circuit Design: Scalable Analog Circuit Design, High-Speed D/A Converters, RF Power Amplifiers*, J. H. Huijsing, M. Steyaert, and A. van Roermund, Eds. Kluwer, Dordrecht, the Netherlands, 2002, pp. 189–210.

[88] D. Giotta, P. Pessl, M. Clara, W. Klatzer, and R. Gaggl, "Low-power 14-bit current steering DAC, for ADSL2+/CO applications in 0.13 μm CMOS," in *European Solid-State Circuits Conference*, 2004, pp. 163–166.

[89] M. Clara, W. Klatzer, B. Seger, A. Di Giandomenico, and L. Gori, "A 1.5 V 200 MS/s 13 b 25 mW DAC with randomized nested background calibration in 0.13 μm CMOS," in *International Solid-State Circuits Conference*, 2007, pp. 250–251.

[90] C.-H. Lin, F. M. L. van der Goes, J. R. Westra *et al.*, "A 12 bit 2.9 GS/s DAC with IM3 < −60 dBc beyond 1 GHz in 65 nm CMOS," *IEEE International Journal of Solid State Circuits*, vol. 44, no. 12, pp. 3285–3293, 2009.

[91] R. Reeder, W. Green, and R. Shillito, "Analog-to-digital converter clock optimization: A test engineering perspective," *Analog Dialogue*, vol. 42, no. 2, pp. 1–7, 2008.

[92] P. R. Gray, P. J. Hurst, S. H. Lewis, and R. G. Meyer, *Analysis and Design of Analog Integrated Circuits*, 4th edn. John Wiley & Sons, New York, NY, 2001.

[93] A. P. Brokaw, "A temperature sensor with single resistor set-point programming," *IEEE International Journal of Solid State Circuits*, vol. 31, no. 12, pp. 1908–1915, 1996.

[94] P. R. Holloway and J. Wan, "Emitter area trim scheme for a PTAT current source," US Patent 7 443 226, 2007.

[95] P. R. Holloway and J. Wan, "Self-regulating process-error trimmable PTAT current source," US Patent 7 236 048, 2007.

[96] G. Ge, C. Zhang, G. Hoogzaad, and K. Makinwa, "A single-trim CMOS bandgap reference with a 3σ inaccuracy of $\pm 0.15\%$ from $-40\,°C$ to $125\,°C$," in *International Solid-State Circuits Conference*, 2010, pp. 78–79.

[97] P. Baginski, P. Brokaw, and S. Wurcer, "A complete 18-bit audio D/A converter," in *International Solid-State Circuits Conference*, 1990, pp. 202–203.

[98] U.-K. Moon and B.-S. Song, "Background digital calibration techniques for pipelined ADC's," *IEEE Transactions on Circuits and Systems – Part II*, vol. 44, no. 2, pp. 102–109, 1997.

[99] S.-U. Kwak, B.-S. Song, and K. L. Bacrania, "A 15-b, 5-Msample/s low-spurious CMOS ADC," *IEEE International Journal of Solid State Circuits*, vol. 32, no. 12, pp. 1866–1875, 1997.

[100] H.-S. Lee, D. A. Hodges, and P. R. Gray, "A self-calibrating 15 bit CMOS A/D converter," *IEEE International Journal of Solid State Circuits*, vol. 19, no. 6, pp. 813–819, 1984.

[101] A. Karanicolas, H.-S. Lee, and K. L. Bacrania, "A 15-b 1-Msample/s digitally self-calibrated pipeline ADC," *IEEE International Journal of Solid State Circuits*, vol. 28, no. 12, pp. 1207–1215, 1993.

[102] J. R. Raol, G. Girjia, and J. Singh, *Modelling and Parameter Estimation of Dynamic Systems*. IEE Control Engineering. Institution of Engineering and Technology (IET) London, 2004.

[103] B. C. Levy, *Principles of Signal Detection and Parameter Estimation*. Springer-Verlag, Berlin, Germany, 2009.

[104] A. van den Bos, *Parameter Estimation for Scientists and Engineers*. John Wiley & Sons, New York, NY, 2007.

[105] R. Gregorian and G. C. Temes, *Analog MOS Integrated Circuits for Signal Processing*. John Wiley & Sons, New York, NY, 1986.

[106] B. Murmann and V. Gopinathan, "ISSCC 2011 evening session on data converter breakthrough in retrospect," in *International Solid-State Circuits Conference*, IEEE, 2011.

[107] R. J. van de Plassche and R. E. J. van der Grift, "A high-speed 7 bit A/D converter," *IEEE International Journal of Solid State Circuits*, vol. 14, no. 6, pp. 938–943, 1979.

[108] K. Bult and A. Buchwald, "An embedded 240-mW 10-b 50-MS/s CMOS ADC in 1-mm^2," *IEEE International Journal of Solid State Circuits*, vol. 32, no. 12, pp. 1887–1895, 1997.

[109] K. Kattmann and J. Barrow, "A technique for reducing differential non-linearity errors in flash A/D converters," in *International Solid-State Circuits Conference*, 1991, pp. 170–171.

[110] H. Pan and A. A. Abidi, "Spatial filtering in flash A/D converters," *IEEE Transactions on Circuits and Systems – Part II*, vol. 50, no. 8, pp. 424–436, 2003.

[111] K. Nagaraj, D. A. Martin, M. Wolfe *et al.*, "A dual-mode 700-Msamples/s 6-bit 200-Msamples/s 7-bit A/D converter in a 0.25-μm digital CMOS process," *IEEE International Journal of Solid State Circuits*, vol. 35, no. 12, pp. 1760–1768, 2000.

[112] M. Choi and A. A. Abidi, "A 6-b 1.3-Gsample/s A/D converter in 0.35-μm CMOS," *IEEE International Journal of Solid State Circuits*, vol. 36, no. 12, pp. 1847–1858, 2001.

[113] P. C. S. Scholtens and M. Vertregt, "A 6-b 1.6-Gsample/s flash ADC in 0.18-μm CMOS using averaging termination," *IEEE International Journal of Solid State Circuits*, vol. 37, no. 12, pp. 1599–1609, 2002.

[114] Y.-I. Park, S. Karthikeyan, F. Tsay, and E. Bartolome, "A 10 b 100 MSample/s CMOS pipelined ADC with 1.8 V power supply," in *International Solid-State Circuits Conference*, 2001, pp. 580–583.

[115] K. Poulton, R. Neff, A. Muto *et al.*, "A 4 GSample/s 8 b ADC in 0.35 μm CMOS," in *International Solid-State Circuits Conference*, 2002, pp. 166–167.

[116] K. Poulton, R. Neff, B. Setterberg *et al.*, "A 20 GS/s 8 b ADC with a 1 MB memory in 0.18 μm CMOS," in *International Solid-State Circuits Conference*, 2003, pp. 318–319.

[117] G. Manganaro, "Feed-forward approach for timing skew in interleaved and double-sampled circuits," *IEE Electronics Letters*, vol. 37, no. 9, pp. 552–553, 2001.

[118] Y. Li and E. Sánchez-Sinencio, "A wide input bandwidth 7-bit 300-MSample/s folding and current-mode interpolating ADC," *IEEE International Journal of Solid State Circuits*, vol. 38, no. 8, pp. 1405–1410, 2003.

[119] G. Geelen and E. Paulus, "An 8 b 600 MS/s 200 mW CMOS folding A/D converter using an amplifier preset technique," in *International Solid-State Circuits Conference*, 2004, pp. 254–255.

[120] R. C. Taft, C. A. Menkus, M. R. Tursi, O. Hidri, and V. Pons, "Advances in high-speed ADC architectures using offset calibration," in *Analog Circuit Design: High-Speed A-D Converters, Automotive Electronics and Ultra-Low Power Wireless*, A. H. M. van Roermund, H. Casier, and M. Steyaert, Eds. Kluwer, Dordrecht, the Netherlands, 2006, pp. 189–210.

[121] P. Vorenkamp and R. Roovers, "A 12-b, 60-MSample/s cascaded folding and interpolating ADC," *IEEE International Journal of Solid State Circuits*, vol. 32, no. 12, pp. 1876–1886, 1997.

[122] M.-J. Choe, B.-S. Song, and K. Bacrania, "An 8-b 100-MSample/s CMOS pipelined folding ADC," *IEEE International Journal of Solid State Circuits*, vol. 36, no. 2, pp. 184–194, 2001.

[123] Y. Nakajima, A. Sakaguchi, T. Ohkido, T. Matsumoto, and M. Yotsuyanagi, "A self-background calibrated 6 b 2.7 GS/s ADC with cascade-calibrated folding-interpolating architecture," in *VLSI Circuits Conference*, 2007, pp. 266–267.

[124] Y. Nakajima, A. Sakaguchi, T. Ohkido, T. Matsumoto, and M. Yotsuyanagi, "A background self-calibrated 6 b 2.7 GS/s ADC with cascade-calibrated folding-interpolating architecture," *IEEE International Journal of Solid State Circuits*, vol. 45, no. 4, pp. 707–718, 2010.

[125] I. Bogue and M. P. Flynn, "A 57 dB SFDR digitally calibrated 500 MS/s folding ADC in 0.18 μm digital CMOS," in *Custom Integrated Circuits Conference*, 2007, pp. 337–340.

[126] K. Makigawa, K. Ono, T. Ohkawa, K. Matsuura, and M. Segami, "A 7 bit 800 Msps 120 mW folding and interpolation ADC using a mixed-averaging scheme," in *VLSI Circuits Conference*, 2006, pp. 138–139.

[127] A. Razzaghi, S.-W. Tam, P. Kalkhoran *et al.*, "A single-channel 10 b 1 GS/s ADC with 1-cycle latency using pipelined cascaded folding," in *Proceedings of IEEE Bipolar/BiCMOS Circuits and Technology Meeting*, 2008, pp. 265–266.

[128] Y. Chen, Q. Huang, and T. Burger, "A 1.2 V 200-MS/s 10-bit folding and interpolating ADC in 0.13-μm CMOS," in *European Solid-State Circuits Conference*, 2007, pp. 155–158.

[129] R. C. Taft, P. A. Francese, M. R. Tursi *et al.*, "A 1.8 V 1.0 GS/s 10 b self-calibrating unified-folding-interpolating ADC with 9.1 ENOB at Nyquist frequency," *IEEE International Journal of Solid State Circuits*, vol. 44, no. 12, pp. 3294–3304, 2009.

[130] G. Van der Plas, S. Decoutere, and S. Donnay, "A 0.16 pJ/conversion-step 2.5 mW 1.25 GS/s 4 b ADC in a 90 nm digital CMOS process," in *International Solid-State Circuits Conference*, 2006, pp. 566–567.

[131] N. Sayiner, H. Sorensen, and T. Viswanathan, "A new signal acquisition technique," in *IEEE International Symposium on Circuits and Systems*, 1992, pp. 1140–1143.

[132] N. Sayiner, H. Sorensen, and T. Viswanathan, "A level-crossing sampling scheme for A/D conversion," *IEEE Transactions on Circuits and Systems – Part II*, vol. 43, no. 4, pp. 335–339, 1996.

[133] Y. Tsividis, "Event-driven data acquisition and digital signal processing – a tutorial," *IEEE Transactions on Circuits and Systems – Part II*, vol. 57, no. 8, pp. 577–581, 2010.

[134] Y. Tsividis, "Event-driven data acquisition and continuous-time digital signal processing," in *Custom Integrated Circuits Conference*, 2010, pp. 1–8.

[135] Y. Tsividis, "Event-driven, continuous-time ADCs and DSPs for adapting power dissipation to signal activity," in *IEEE International Symposium on Circuits and Systems*, 2010, pp. 3581–3584.

[136] C. S. Burrus, R. A. Gopinath, and H. Guo, *Introduction to Wavelets and Wavelet Transforms: A Primer*. Prentice Hall, New York, NY, 1998.

[137] G. Russell, D. J. Kinniment, E. G. Chester, and M. R. McLauchlan, *CAD for VLSI*. Van Nostrand Reinhold, London, UK, 1985.

[138] J. K. White and A. Sangiovanni-Vincentelli, *Relaxation Techniques for the Simulation of VLSI Circuits*. Kluwer, Dordrecht, the Netherlands, 1987.

[139] P. H. Ellis, "Extension of phase plane analysis to quantized systems," *IRE Transactions on Automatic Control*, vol. 4, no. 2, pp. 43–54, 1959.

[140] R. C. Dorf, M. C. Farren, and C. A. Phillips, "Adaptive sampling frequency for sampled-data control systems," *IRE Transactions on Automatic Control*, vol. 7, no. 1, pp. 38–47, 1962.

[141] D. Donoho, "Compressed sensing," *IEEE Transactions on Information Theory*, vol. 52, no. 4, pp. 1289–1306, 2006.

[142] M. Trakimas and S. R. Sonkusale, "An adaptive resolution asynchronous ADC architecture for data compression in energy constrained sensing applications," *IEEE Transactions on Circuits and Systems – Part I*, vol. 58, 2011.

[143] F. Akopyan, R. Manohar, and A. B. Apsel, "A level-crossing flash asynchronous analog-to-digital converter," in *IEEE International Symposium on Asynchronous Circuits and Systems*, 2006, pp. 12–22.

[144] B. Schell and Y. Tsividis, "A continuous-time ADC/DSP/DAC system with no clock and with activity-dependent power dissipation," *IEEE International Journal of Solid State Circuits*, vol. 43, no. 11, pp. 2472–2481, 2008.

[145] M. Trakimas and S. R. Sonkusale, "A 0.8 V asynchronous ADC for energy constrained sensing applications," in *Custom Integrated Circuits Conference*, 2008, pp. 173–176.

[146] M. Kurchuk and Y. Tsividis, "Signal-dependent variable-resolution clockless A/D conversion with application to continuous-time digital signal processing," *IEEE Transactions on Circuits and Systems – Part I*, vol. 57, no. 5, pp. 982–991, 2010.

[147] J. C. Schelleng, "Code modulation communication system," US Patent 2 453 461, 1948.

[148] B. M. Gordon and E. T. Colton, "Signal conversion apparatus," US Patent 2 997 704, 1961.

[149] A. Baschirotto, "ISSCC 2009 tutorial succesive approximation register (SAR) A/D converters," in *International Solid-State Circuits Conference*, IEEE, 2009.

[150] J. L. McCreary and P. R. Gray, "All-MOS charge redistribution analog-to-digital conversion techniques, part I," *IEEE International Journal of Solid State Circuits*, vol. 10, no. 6, pp. 371–379, 1975.

[151] F. Kuttner, "A 1.2 V 10 b 20 MSample/s non-binary successive approximation ADC in 0.13 μm CMOS," in *International Solid-State Circuits Conference*, 2002, pp. 136–137.

[152] P. Confalonieri, M. Zamprogno, F. Girardi, G. Nicollini, and A. Nagari, "A 2.7 mW 1 MSPS 10 b analog-to-digital converter with built-in reference buffer and 1 LSB accuracy programmable input ranges," in *European Solid-State Circuits Conference*, 2004, pp. 255–258.

[153] J. Craninckx and G. Van der Plas, "A 65 fJ/conversion-step 0-to-50 MS/s 0-to-0.7 mW 9 b charge-sharing SAR ADC in 90 nm digital CMOS," in *International Solid-State Circuits Conference*, 2007, pp. 246–247.

[154] G. Manganaro, S.-U. Kwak, and A. Bugeja, "A dual 10-b 200-MSPS pipelined D/A converter with DLL-based clock synthesizer," *IEEE International Journal of Solid State Circuits*, vol. 39, no. 11, pp. 1829–1838, 2004.

[155] S. H. Lewis, H. S. Fetterman, G. F. Gross Jr., R. Ramachandran, and T. R. Viswanathan, "A 10-b 20-Msample/s analog-to-digital converter," *IEEE International Journal of Solid State Circuits*, vol. 27, no. 3, pp. 351–358, 1992.

[156] V. Giannini, P. Nuzzo, V. Chironi *et al.*, "An 820 μW 9 b 40 MS/s noise-tolerant dynamic-SAR ADC in 90 nm digital CMOS," in *International Solid-State Circuits Conference*, 2008, pp. 238–239.

[157] S.-W. M. Chen and R. W. Brodersen, "A 6-bit 600-MS/s 5.3-mW asynchronous ADC in 0.13-μm CMOS," *IEEE International Journal of Solid State Circuits*, vol. 41, no. 12, pp. 2669–2680, 2006.

[158] B. P. Ginsburg and A. P. Chandrakasan, "An energy-efficient charge recycling approach for a SAR converter with capacitive DAC," in *IEEE International Symposium on Circuits and Systems*, vol. 1, 2005, pp. 184–187.

[159] B. P. Ginsburg and A. P. Chandrakasan, "500-MS/s 5-bit ADC in 65-nm CMOS with split capacitor array DAC," *IEEE International Journal of Solid State Circuits*, vol. 42, no. 4, pp. 739–747, 2007.

[160] M. van Elzakker, E. van Tuijl, P. Geraedts *et al.*, "A 10-bit charge-redistribution ADC consuming 1.9 μW at 1 MS/s," *IEEE International Journal of Solid State Circuits*, vol. 45, no. 5, pp. 1007–1015, 2010.

[161] L. J. Svensson and J. G. Koller, "Driving a capacitive load without dissipating fCV^2," in *IEEE Symposium on Low Power Electronics*, 1994, pp. 100–101.

[162] D. Draxelmayr, "A 6 b 600 MHz 10 mW ADC array in digital 90 nm CMOS," in *International Solid-State Circuits Conference*, 2004, pp. 264–265.

[163] B. P. Ginsburg and A. P. Chandrakasan, "Highly interleaved 5-bit, 250-MSample/s, 1.2-mW ADC with redundant channels in 65-nm CMOS," *IEEE International Journal of Solid State Circuits*, vol. 43, no. 12, pp. 2641–2650, 2008.

[164] J. McNeill, M. C. W. Coln, and B. J. Larivee, " 'Split ADC' architecture for deterministic digital background calibration of a 16-bit 1-MS/s ADC," *IEEE International Journal of Solid State Circuits*, vol. 40, no. 12, pp. 2437–2445, 2005.

[165] G. Van der Plas and B. Verbruggen, "A 150 MS/s 133 μW 7 bit ADC in 90 nm digital CMOS," *IEEE International Journal of Solid State Circuits*, vol. 43, no. 12, pp. 2631–2640, 2008.

[166] C. P. Hurrell, C. Lyden, D. Laing, D. Hummerston, and M. Vickery, "An 18 b 12.5 MS/s ADC with 93 dB SNR," *IEEE International Journal of Solid State Circuits*, vol. 45, no. 12, pp. 2647–2654, 2010.

[167] H. Wei, C.-H. Chan, U.-F. Chio *et al.*, "A 0.024 mm² 8 b 400 MS/s SAR ADC with 2 b/cycle and resistive DAC in 65 nm CMOS," in *International Solid-State Circuits Conference*, 2011, pp. 187–188.

[168] E. M. Deloraine, S. Van Mierlo, and B. Derjavitch, "Méthode et système de transmission par impulsions," French Patent 932 140, 1946.

[169] E. M. Deloraine, S. Van Mierlo, and B. Derjavitch, "Communication system utilizing constant amplitude pulses of opposite polarities," US Patent 2 629 857, 1953.

[170] F. de Jager, "Delta modulation: A method of PCM transmission using the one unit code," *Phillips Research Report*, vol. 7, pp. 542–546, 1952.

[171] H. Van de Weg, "Quantizing noise of a single integration delta modulation system with an *N*-digit code," *Phillips Research Report*, vol. 8, pp. 367–385, 1953.

[172] C. C. Cutler, "Differential quantization of communication signals," US Patent 2 605 361, 1952.

[173] C. C. Cutler, "Transmission systems employing quantization," US Patent 2 927 962, 1960.

[174] H. Inose, Y. Yasuda, and J. Murakami, "A telemetering system by code modulation – Δ–Σ modulation," *IRE Transactions on Space Electronics and Telemetry*, vol. 8, no. 9, pp. 204–209, 1962.

[175] S. R. Norsworthy, R. Schreier, and G. C. Temes, Eds., *Delta–Sigma Data Converters: Theory, Design and Simulation*. IEEE Press, Piscataway, NJ, 1996.

[176] S. Pavan, "Alias rejection of continuous-time $\Delta\Sigma$ modulators with switched-capacitor feedback DACs," *IEEE Transactions on Circuits and Systems – Part I*, vol. 58, no. 2, pp. 233–243, 2011.

[177] M. Bolatkale, L. J. Breems, R. Rutten, and K. A. Makinwa, "A 4 GHz CT ADC with 70 dB DR and 74 dBFS THD in 125 MHz BW," in *International Solid-State Circuits Conference*, 2011, pp. 470–471.

[178] G. Mitteregger, C. Ebner, S. Mechnig *et al.*, "A 20-mW 640-MHz CMOS continuous-time $\Sigma\Delta$ ADC with 20-MHz signal bandwidth, 80-dB dynamic range and 12-bit ENOB," *IEEE International Journal of Solid State Circuits*, vol. 41, no. 12, pp. 2641–2649, 2006.

[179] R. H. van Veldhoven, R. Rutten, and L. J. Breems, "An inverter-based hybrid $\Delta\Sigma$ modulator," in *International Solid-State Circuits Conference*, 2008, pp. 492–493.

[180] S. Patón, A. Di Giandomenico, L. Hernández *et al.*, "A 70-mW 300-MHz CMOS continuous-time $\Sigma\Delta$ ADC with 15-MHz bandwidth and 11 bits of resolution," *IEEE International Journal of Solid State Circuits*, vol. 39, no. 7, pp. 1056–1063, 2004.

[181] M. Keller, A. Buhmann, J. Sauerbrey, M. Ortmanns, and Y. Manoli, "A comparative study on excess-loop-delay compensation techniques for continuous-time sigma–delta modulators," *IEEE Transactions on Circuits and Systems – Part I*, vol. 55, no. 11, pp. 3480–3487, 2008.

[182] K. Nguyen, R. Adams, K. Sweetland, and H. Chen, "A 106-dB SNR hybrid oversampling analog-to-digital converter for digital audio," *IEEE International Journal of Solid State Circuits*, vol. 40, no. 12, pp. 2408–2415, 2005.

[183] P. Morrow, M. Chamarro, C. Lyden *et al.*, "A 0.18 μm 102 dB-SNR mixed CT SC audio-band ADC," in *International Solid-State Circuits Conference*, 2005, pp. 178–179.

[184] L. Dörrer, F. Kuttner, A. Santner *et al.*, "A 2.2 mW, continuous-time Sigma–Delta ADC for voice coding with 95 dB dynamic range in a 65 nm CMOS process," in *European Solid-State Circuits Conference*, 2006, pp. 195–198.

[185] L. Dörrer, F. Kuttner, A. Santner *et al.*, "A continuous time $\Delta\Sigma$ ADC for voice coding with 92 dB DR in 45 nm CMOS," in *International Solid-State Circuits Conference*, 2008, pp. 502–503.

[186] I. Galton, "Delta–sigma data conversion in wireless transceivers," *IEEE Transactions on Microwave Theory and Techniques*, vol. 50, no. 1, pp. 303–315, 2002.

[187] G. Raghavan, J. F. Jensen, J. Laskowski *et al.*, "Architecture, design, and test of continuous-time tunable intermediate-frequency bandpass delta–sigma modulators," *IEEE International Journal of Solid State Circuits*, vol. 36, no. 1, pp. 5–13, 2001.

[188] R. Schreier, J. Lloyd, L. Singer *et al.*, "A 10–300-MHz IF-digitizing IC with 90–105-dB dynamic range and 15–333-kHz bandwidth," *IEEE International Journal of Solid State Circuits*, vol. 37, no. 12, pp. 1636–1644, 2002.

[189] R. Schreier, N. Abaskharoun, H. Shibata *et al.*, "A 375-mW quadrature bandpass ΔΣ ADC with 8.5-MHz BW and 90-dB DR at 44 MHz," *IEEE International Journal of Solid State Circuits*, vol. 41, no. 12, pp. 2632–2640, 2006.

[190] L. J. Breems, R. Rutten, R. H. M. van Veldhoven, and G. van der Weide, "A 56 mW continuous-time quadrature cascaded ΔΣ modulator with 77 dB DR in a near zero-IF 20 MHz band," *IEEE International Journal of Solid State Circuits*, vol. 42, no. 12, pp. 2696–2705, 2007.

[191] P. G. R. Silva, L. J. Breems, K. A. A. Makinwa, M. Raf Roovers, and J. H. Huijsing, "An IF-to-baseband ΣΔ modulator for AM/FM/IBOC radio receivers with a 118 dB dynamic range," *IEEE International Journal of Solid State Circuits*, vol. 42, no. 5, pp. 1076–1089, 2007.

[192] B. K. Thandri and J. Silva-Martinez, "A 63 dB SNR, 75-mW bandpass RF ΣΔ ADC at 950 MHz using 3.8-GHz clock in 0.25-μm SiGe BiCMOS technology," *IEEE International Journal of Solid State Circuits*, vol. 42, no. 2, pp. 269–279, 2007.

[193] W. Yang, W. Schofield, H. Shibata *et al.*, "A 100 mW 10 MHz-BW CT modulator with 87 dB DR and 91 dBc IMD," in *International Solid-State Circuits Conference*, 2008, pp. 498–499.

[194] L. R. Carley, "A noise-shaping coder topology for 15+ bit converters," *IEEE International Journal of Solid State Circuits*, vol. 24, no. 2, pp. 267–273, 1989.

[195] R. T. Baird and T. S. Fiez, "Linearity enhancement of multibit ΔΣ A/D and D/A converters using data weighted averaging," *IEEE Transactions on Circuits and Systems – Part II*, vol. 42, no. 12, pp. 753–762, 1995.

[196] R. E. Radke, A. Eshraghi, and T. S. Fiez, "A 14-bit current-mode ΔΣ DAC based upon rotated data weighted averaging," *IEEE International Journal of Solid State Circuits*, vol. 35, no. 8, pp. 1074–1084, 2000.

[197] F. Chen and B. H. Leung, "A high resolution multibit sigma–delta modulator with individual level averaging," *IEEE International Journal of Solid State Circuits*, vol. 30, no. 4, pp. 453–460, 1995.

[198] A. A. Hamoui and K. W. Martin, "High-order multibit modulators and pseudo data-weighted-averaging in low-oversampling ΔΣ ADCs for broad-band applications," *IEEE Transactions on Circuits and Systems – Part I*, vol. 51, no. 1, pp. 72–85, 2004.

[199] D.-H. Lee, Y.-H. Lin, and T.-H. Kuo, "Nyquist-rate current-steering digital-to-analog converters with random multiple data-weighted averaging technique and Q^N rotated walk switching scheme," *IEEE Transactions on Circuits and Systems – Part II*, vol. 53, no. 11, pp. 1264–1268, 2006.

[200] S.-J. Huang and Y.-Y. Lin, "A 1.2 V 2 MHz BW 0.084 mm^2 CT ΔΣ ADC with −97.7 dBc THD and 80 dB DR using low-latency DEM," in *International Solid-State Circuits Conference*, 2009, pp. 172–173.

[201] J. Ryckaert, J. Borremans, B. Verbruggen *et al.*, "A 2.4 GHz low-power sixth-order RF bandpass ΔΣ converter in CMOS," *IEEE International Journal of Solid State Circuits*, vol. 44, no. 11, pp. 2873–2880, 2009.

[202] I. Galdi, E. Bonizzoni, P. Malcovati, G. Manganaro, and F. Maloberti, "40 MHz IF 1 MHz bandwidth two-path bandpass ΣΔ modulator with 72 dB DR consuming 16 mW," *IEEE International Journal of Solid State Circuits*, vol. 43, no. 7, pp. 1648–1656, 2008.

[203] C. Lyden, J. Ryan, C. A. Ugarte, J. Kornblum, and F. M. Yung, "A single shot sigma delta analog to digital converter for multiplexed applications," in *Custom Integrated Circuits Conference*, 1995, pp. 203–206.

[204] P. R. Holloway, E. D. Blom, J. Wan, and S. H. Urie, "Digitizing temperature measurement system and method of operation," US Patent 6 962 436, 2005.

[205] C. Jansson, "A high-resolution, compact, and low-power ADC suitable for array implementation in standard CMOS," *IEEE Transactions on Circuits and Systems – Part I*, vol. 42, no. 11, pp. 904–912, 1995.

[206] R. Harjani and T. A. Lee, "FRC: A method for extending the resolution of Nyquist rate converters using oversampling," *IEEE Transactions on Circuits and Systems – Part II*, vol. 45, no. 4, pp. 482–494, 1998.

[207] P. Rombouts, W. D. Wilde, and L. Weyten, "A 13.5-b 1.2-V micropower extended counting A/D converter," *IEEE International Journal of Solid State Circuits*, vol. 36, no. 2, pp. 176–183, 2001.

[208] C. Schott, R. Racz, A. Manco, and N. Simonne, "CMOS single-chip electronic compass with microcontroller," *IEEE International Journal of Solid State Circuits*, vol. 42, no. 12, pp. 2923–2933, 2007.

[209] H.-S. Lee, L. Brooks, and C. G. Sodini, "Zero-crossing-based ultra-low-power A/D converters," *Proceedings of the IEEE*, vol. 98, no. 2, pp. 315–332, 2010.

[210] J. K. Fiorenza, T. Sepke, P. Holloway, C. G. Sodini, and H.-S. Lee, "Comparator-based switched-capacitor circuits for scaled CMOS technologies," *IEEE International Journal of Solid State Circuits*, vol. 41, no. 12, pp. 2658–2668, 2006.

[211] L. Brooks and H.-S. Lee, "A zero-crossing-based 8-bit 200 MS/s pipelined ADC," *IEEE International Journal of Solid State Circuits*, vol. 42, no. 12, pp. 2677–2687, 2007.

[212] S.-K. Shin, Y.-S. You, S.-H. Lee *et al.*, "A fully-differential zero-crossing-based 1.2 V 10 b 26 MS/s pipelined ADC in 65 nm CMOS," in *VLSI Circuits Conference*, 2008, pp. 218–219.

[213] T. Musah, S. Kwon, H. Lakdawala, K. Soumyanath, and U.-K. Moon, "A 630 μW zero-crossing-based ΔΣ ADC using switched-resistor current sources in 45 nm CMOS," in *Custom Integrated Circuits Conference*, 2009, pp. 1–4.

[214] L. Brooks and H.-S. Lee, "A 12 b, 50 MS/s, fully differential zero-crossing based pipelined ADC," *IEEE International Journal of Solid State Circuits*, vol. 44, no. 12, pp. 3329–3343, 2009.

[215] L. Brooks and H.-S. Lee, "Background calibration of pipelined ADCs via decision boundary gap estimation," *IEEE Transactions on Circuits and Systems – Part I*, vol. 55, no. 10, pp. 2969–2979, 2008.

[216] T. Sepke, P. Holloway, C. G. Sodini, and H.-S. Lee, "Noise analysis for comparator-based circuits," *IEEE Transactions on Circuits and Systems – Part I*, vol. 56, no. 3, pp. 541–553, 2009.

[217] B. R. Gregoire and U.-K. Moon, "An over-60 dB true rail-to-rail performance using correlated level shifting and an opamp with only 30 dB loop gain," *IEEE International Journal of Solid State Circuits*, vol. 43, no. 12, pp. 2620–2630, 2008.

[218] K. Nagaraj, "Switched-capacitor circuits with reduced sensitivity to amplifier gain," *IEEE Transactions on Circuits and Systems*, vol. 34, no. 5, pp. 571–574, 1987.

[219] B. Hershberg, S. Weaver, and U.-K. Moon, "Design of a split-CLS pipelined ADC with full signal swing using an accurate but fractional signal swing opamp," *IEEE International Journal of Solid State Circuits*, vol. 45, no. 12, pp. 2623–2633, 2010.

[220] B. R. Veillette and G. W. Roberts, "High frequency sinusoidal generation using delta–sigma modulation techniques," in *IEEE International Symposium on Circuits and Systems*, 1995, pp. 637–640.

[221] E. Roza, "Analog-to-digital conversion via duty-cycle modulation," *IEEE Transactions on Circuits and Systems – Part II*, vol. 44, no. 11, pp. 907–914, 1997.

[222] D. Santos, S. Dow, J. Flasck, and M. Levi, "A CMOS delay locked loop and sub-nanosecond time-to-digital converter chip," *IEEE Transactions on Nuclear Science*, vol. 43, no. 3, pp. 1717–1727, 1996.

[223] G. W. Roberts and M. Ali-Bakhshian, "A brief introduction to time-to-digital and digital-to-time converters," *IEEE Transactions on Circuits and Systems – Part II*, vol. 57, no. 3, pp. 153–157, 2010.

[224] R. E. Best, Ed., *Phase-Locked Loops: Design, Simulation and Applications*, 4th edn. McGraw-Hill, New York, NY, 1999.

[225] P. Dudek, S. Szczepański, and J. V. Hatfield, "A high-resolution CMOS time-to-digital converter utilizing a Vernier delay line," *IEEE International Journal of Solid State Circuits*, vol. 35, no. 2, pp. 240–247, 2000.

[226] S. Sidiropoulos and M. A. Horowitz, "A semidigital dual delay-locked loop," *IEEE International Journal of Solid State Circuits*, vol. 32, no. 11, pp. 1683–1692, 1997.

[227] P. K. Hanumolu, V. Kratyuk, G.-Y. Wei, and U.-K. Moon, "A sub-picosecond resolution 0.5–1.5 GHz digital-to-phase converter," *IEEE International Journal of Solid State Circuits*, vol. 43, no. 2, pp. 414–424, 2008.

[228] A. Iwata, N. Sakimura, M. Nagata, and T. Morie, "The architecture of delta sigma analog-to-digital converters using a voltage-controlled oscillator as a multibit quantizer," *IEEE Transactions on Circuits and Systems – Part II*, vol. 46, no. 7, pp. 941–945, 1999.

[229] Y.-G. Yoon, J. Kim, T.-K. Jang, and S. Cho, "A time-based bandpass ADC using time-interleaved voltage-controlled oscillators," *IEEE Transactions on Circuits and Systems – Part I*, vol. 55, no. 11, pp. 3571–3581, 2008.

[230] M. Høvin, A. Olsen, T. S. Lande, and C. Toumazou, "A mostly-digital variable-rate continuous-time delta–sigma modulator ADC," *IEEE International Journal of Solid State Circuits*, vol. 32, no. 1, pp. 13–22, 1997.

[231] M. Z. Straayer and M. H. Perrott, "A 12-bit, 10-MHz bandwidth, continuous-time $\Sigma\Delta$ ADC with a 5-bit, 950-MS/s VCO-based quantizer," *IEEE International Journal of Solid State Circuits*, vol. 43, no. 4, pp. 805–814, 2008.

[232] C. Wulff and T. Ytterdal, "Resonators in open-loop sigma–delta modulators," *IEEE Transactions on Circuits and Systems – Part I*, vol. 56, no. 10, pp. 2159–2172, 2009.

[233] Y.-G. Yoon and S. Cho, "A 1.5-GHz 63 dB SNR 20 mW direct RF sampling bandpass VCO-based ADC in 65 nm CMOS," in *VLSI Circuits Conference*, 2009, pp. 270–271.

[234] J. Kim, T.-K. Jang, Y.-G. Yoon, and S. Cho, "Analysis and design of voltage-controlled oscillator based analog-to-digital converter," *IEEE Transactions on Circuits and Systems – Part I*, vol. 57, no. 1, pp. 18–30, 2010.

[235] G. Taylor and I. Galton, "A mostly-digital variable-rate continuous-time delta–sigma modulator ADC," *IEEE International Journal of Solid State Circuits*, vol. 45, no. 12, pp. 2634–2646, 2010.

[236] M. Park and M. H. Perrott, "A 78 dB SNDR 87 mW 20 MHz bandwidth continuous-time $\Delta\Sigma$ ADC with VCO-based integrator and quantizer implemented in 0.13 μ CMOS," *IEEE International Journal of Solid State Circuits*, vol. 44, no. 12, pp. 3344–3358, 2009.

[237] A. Agnes, E. Bonizzoni, P. Malcovati, and F. Maloberti, "A 9.4-ENOB 1V 3.8 μW 100 kS/s SAR ADC with time-domain comparator," in *International Solid-State Circuits Conference*, 2008, pp. 246–247.

[238] S. Naraghi, M. Courcy, and M. P. Flynn, "A 9-bit, 14 μw and 0.06 mm^2 pulse position modulation ADC in 90 nm digital CMOS," *IEEE International Journal of Solid State Circuits*, vol. 45, no. 9, pp. 1870–1880, 2010.

[239] D. Robertson, "The past, present, and future of data converters and mixed signal ICs: A 'universal' model," in *VLSI Circuits Conference*, 2006, pp. 1–4.

[240] B. Murmann and B. E. Boser, "A 12-bit 75-MS/s pipelined ADC using open-loop residue amplification," *IEEE International Journal of Solid State Circuits*, vol. 38, no. 12, pp. 2040–2050, 2003.

[241] W. Yang, D. Kelly, I. Mehr, M. T. Sayuk, and L. Singer, "A 3-V 340-mW 14-b 75-Msample/s CMOS ADC with 85-dB SFDR at Nyquist input," *IEEE International Journal of Solid State Circuits*, vol. 36, no. 12, pp. 1931–1936, 2001.

[242] A. A. Abidi, A. Matsuzawa, B. Staszewski *et al.*, "A-SSCC 2008 panel on digitally assisted analog and RF circuits: Potentials and issues," in *IEEE Asian Solid-State Circuits Conference*, IEEE, 2008.

[243] A. Baschirotto, "ISSCC 2008 forum on digitally-assisted analog & RF circuits," in *International Solid-State Circuits Conference*, IEEE, 2008.

[244] J. Li and U.-K. Moon, "Background calibration techniques for multistage pipelined ADCs with digital redundancy," *IEEE Transactions on Circuits and Systems – Part II*, vol. 50, no. 9, pp. 531–538, 2003.

[245] E. Siragusa and I. Galton, "A digitally enhanced 1.8-V 15-bit 40-MSample/s CMOS pipelined ADC," *IEEE International Journal of Solid State Circuits*, vol. 39, no. 12, pp. 2126–2138, 2004.

[246] Y. Chiu, C. W. Tsang, B. Nikolić, and P. R. Gray, "Least mean square adaptive digital background calibration of pipelined analog-to-digital converters," *IEEE Transactions on Circuits and Systems – Part I*, vol. 51, no. 1, pp. 38–46, 2004.

[247] J. P. Keane, P. J. Hurst, and S. H. Lewis, "Background interstage gain calibration technique for pipelined ADCs," *IEEE Transactions on Circuits and Systems – Part I*, vol. 52, no. 1, pp. 32–43, 2005.

[248] J. P. Keane, P. J. Hurst, and S. H. Lewis, "Digital background calibration for memory effects in pipelined analog-to-digital converters," *IEEE Transactions on Circuits and Systems – Part I*, vol. 53, no. 3, pp. 511–525, 2006.

[249] S. Heinen and D. Leenaerts, "ISSCC 2010 forum on reconfigurable RF and data converters," in *International Solid-State Circuits Conference*, IEEE, 2010.

[250] P. Crombez, G. Van der Plas, M. S. J. Steyaert, and J. Craninckx, "A single-bit 500 kHz–10 MHz multimode power-performance scalable 83-to-67 dB DR CT ΔΣ for SDR in 90 nm digital CMOS," *IEEE International Journal of Solid State Circuits*, vol. 45, no. 6, pp. 1159–1171, 2010.

[251] P. Malla, H. Lakdawala, K. Kornegay, and K. Soumyanath, "A 28 mW spectrum-sensing reconfigurable 20 MHz 72 dB-SNR 70 dB-SNDR DT ADC for 802.11n/WiMAX receivers," in *International Solid-State Circuits Conference*, 2008, pp. 496–497.

[252] Y. Oh and B. Murmann, "A low-power, 6-bit time-interleaved SAR ADC using OFDM pilot tone calibration," in *Custom Integrated Circuits Conference*, 2007, pp. 193–196.

[253] T. L. Brooks, D. H. Robertson, D. F. Kelly, A. Del Muro, and S. W. Harston, "A cascaded sigma–delta pipeline A/D converter with 1.25 MHz signal bandwidth and 89 dB

SNR," *IEEE International Journal of Solid State Circuits*, vol. 32, no. 12, pp. 1896–1906, 1997.

[254] B. Verbruggen, J. Craninckx, M. Kuijk, P. Wambacq, and G. Van der Plas, "A 2.2 mW 5 b 1.75 GS/s folding flash ADC in 90 nm digital CMOS," in *International Solid-State Circuits Conference*, 2008, pp. 252–253.

[255] R. Verbruggen, J. Craninckx, M. Kuijk, P. Wambacq, and G. Van der Plas, "A 2.6 mW 6 b 2.2 GS/s 4-times interleaved fully dynamic pipelined ADC in 40 nm digital CMOS," in *International Solid-State Circuits Conference*, 2010, pp. 296–297.

[256] W. C. Black Jr. and D. A. Hodges, "Time interleaved converter arrays," *IEEE International Journal of Solid State Circuits*, vol. 15, no. 6, pp. 1022–1029, 1980.

[257] K. Doris, E. Janssen, C. Nani, A. Zanikopoulos, and G. Van der Weide, "A 480 mW 2.6 GS/s 10 b 65 nm CMOS time-interleaved ADC with 48.5 db SNDR up to Nyquist," in *International Solid-State Circuits Conference*, 2011, pp. 180–181.

[258] A. Petraglia and S. K. Mitra, "Analysis of mismatch effects among A/D converters in a time-interleaved waveform digitizer," *IEEE Transactions on Instrumentation and Measurement*, vol. 40, no. 5, pp. 831–835, 1991.

[259] N. Kurosawa, H. Kobayashi, K. Maruyama, H. Sugawara, and K. Kobayashi, "Explicit analysis of channel mismatch effects in time-interleaved ADC systems," *IEEE Transactions on Circuits and Systems – Part I*, vol. 48, no. 3, pp. 261–271, 2001.

[260] C. Vogel, "The impact of combined channel mismatch effects in time-interleaved ADCs," *IEEE Transactions on Instrumentation and Measurement*, vol. 54, no. 1, pp. 415–427, 2005.

[261] D. Fu, K. C. Dyer, S. H. Lewis, and P. J. Hurst, "A digital background calibration technique for time-interleaved analog-to-digital converters," *IEEE International Journal of Solid State Circuits*, vol. 33, no. 12, pp. 1904–1911, 1998.

[262] K. C. Dyer, D. Fu, S. H. Lewis, and P. J. Hurst, "An analog background calibration technique for time-interleaved analog-to-digital converters," *IEEE International Journal of Solid State Circuits*, vol. 33, no. 12, pp. 1912–1919, 1998.

[263] T.-H. Tsai, P. J. Hurst, and S. H. Lewis, "Bandwidth mismatch and its correction in time-interleaved analog-to-digital converters," *IEEE Transactions on Circuits and Systems – Part II*, vol. 53, no. 10, pp. 1133–1137, 2006.

[264] S. M. Jamal, D. Fu, N. C.-J. Chang, P. J. Hurst, and S. H. Lewis, "A 10-b 120-Msample/s time-interleaved analog-to-digital converter with digital background calibration," *IEEE International Journal of Solid State Circuits*, vol. 37, no. 12, pp. 1618–1627, 2002.

[265] J. Mulder, F. M. van der Goes, D. Vecchi *et al.*, "An 800 MS/s dual-residue pipeline ADC in 40 nm CMOS," in *International Solid-State Circuits Conference*, 2011, pp. 184–185.

[266] K. Y. Kim, N. Kusayanagi, and A. A. Abidi, "A 10-b, 100-MS/s CMOS A/D converter," *IEEE International Journal of Solid State Circuits*, vol. 32, no. 3, pp. 302–311, 1997.

[267] H. Jin and E. K. F. Lee, "A digital-background calibration technique for minimizing timing-error effects in time-interleaved ADCs," *IEEE Transactions on Circuits and Systems – Part II*, vol. 47, no. 7, pp. 603–613, 2000.

[268] D. Camarero, K. B. Kalaia, J.-F. Naviner, and P. Loumeau, "Mixed-signal clock-skew calibration technique for time-interleaved ADCs," *IEEE Transactions on Circuits and Systems – Part I*, vol. 55, no. 11, pp. 3676–3687, 2008.

[269] J. Elbornsson and J.-E. Eklund, "Blind estimation of timing errors in interleaved A/D converters," in *IEEE International Conference on Acoustics, Speech, and Signal Processing*, vol. 6, 2001, pp. 3913–3917.

[270] J. Elbornsson, F. Gustafsson, and J.-E. Eklund, "Blind equalization of time errors in a time-interleaved ADC system," *IEEE Transactions on Signal Processing*, vol. 53, no. 4, pp. 1413–1424, 2005.

[271] M. Waltari and K. Halonen, "Timing skew insensitive switching for double sampled circuits," in *IEEE International Symposium on Circuits and Systems*, vol. 1, 1999, pp. 61–64.

[272] M. Gustavsson and N. N. Tan, "Explicit analysis of channel mismatch effects in time-interleaved ADC systems," *IEEE Transactions on Circuits and Systems – Part II*, vol. 47, no. 9, pp. 821–831, 2000.

[273] G. Manganaro, "Feed-forward approach for timing skew in interleaved and double-sampled circuits," US Patent 6 542 017, 2003.

[274] S. K. Gupta, M. A. Inerfield, and J. Wang, "A 1-GS/s 11-bit ADC with 55-dB SNDR, 250-mW power realized by a high bandwidth scalable time-interleaved architecture," *IEEE International Journal of Solid State Circuits*, vol. 41, no. 12, pp. 2650–2657, 2006.

[275] Y. M. Greshishchev, J. Aguirre, M. Besson *et al.*, "A 40 GS/s 6 b ADC in 65 nm CMOS," in *International Solid-State Circuits Conference*, 2010, pp. 390–391.

[276] K. Poulton, K. L. Knudsen, J. Kerley *et al.*, "An 8-GSa/s 8-bit ADC system," in *VLSI Circuits Conference*, 1997, pp. 23–24.

[277] M. Tamba, A. Shimizu, H. Munakata, and T. Komuro, "A method to improve SFDR with random interleaved sampling method," in *IEEE International Test Conference*, 2001, pp. 512–520.

[278] G. Bernardinis, P. Malcovati, F. Maloberti, and E. Soenen, "Dynamic stage matching for parallel pipeline A/D converters," in *IEEE International Symposium on Circuits and Systems*, vol. 1, 2002, pp. 905–909.

[279] K. El-Sankary, A. Assi, and M. Sawan, "New sampling method to improve the SFDR of time-interleaved ADCs," in *IEEE International Symposium on Circuits and Systems*, vol. 1, 2003, pp. 833–836.

[280] J. Elbornsson, F. Gustafsson, and J.-E. Eklund, "Analysis of mismatch effects in a randomly interleaved A/D converter system," *IEEE Transactions on Circuits and Systems – Part I*, vol. 52, no. 3, pp. 465–476, 2005.

[281] I. Galton, "Digital cancellation of D/A converter noise in pipelined A/D converters," *IEEE Transactions on Circuits and Systems – Part II*, vol. 47, no. 3, pp. 185–196, 2000.

[282] E. J. Siragusa and I. Galton, "Gain error correction technique for pipelined analogue-to-digital converters," *IEE Electronics Letters*, vol. 36, no. 7, pp. 617–618, 2000.

[283] X. Wang, P. J. Hurst, and S. H. Lewis, "A 12-bit 20-Msample/s pipelined analog-to-digital converter with nested digital background calibration," *IEEE International Journal of Solid State Circuits*, vol. 39, no. 9, pp. 1799–1808, 2004.

[284] C. R. Grace, P. J. Hurst, and S. H. Lewis, "A 12-bit 80-MSample/s pipelined ADC with bootstrapped digital calibration," *IEEE International Journal of Solid State Circuits*, vol. 40, no. 5, pp. 1038–1046, 2005.

[285] N. Sun, H.-S. Lee, and D. Ham, "Digital background calibration in pipelined ADCs using commutated feedback capacitor switching," *IEEE Transactions on Circuits and Systems – Part II*, vol. 55, no. 9, pp. 877–881, 2008.

[286] H. Wang, X. Wang, P. J. Hurst, and S. H. Lewis, "Nested digital background calibration of a 12-bit pipelined ADC without an input SHA," *IEEE International Journal of Solid State Circuits*, vol. 44, no. 10, pp. 2780–2789, 2009.

[287] I. Mehr and L. Singer, "A 55-mW, 10-bit, 40-Msample/s Nyquist-rate CMOS ADC," *IEEE International Journal of Solid State Circuits*, vol. 35, no. 3, pp. 318–325, 2000.

[288] D.-Y. Chang, "Design techniques for a pipelined ADC without using a front-end sample-and-hold amplifier," *IEEE Transactions on Circuits and Systems – Part I*, vol. 51, no. 11, pp. 2123–2132, 2004.

[289] M. Yoshioka, M. Kudo, K. Gotoh, and Y. Watanabe, "A 10 b 125 MS/s 40 mW pipelined ADC in 0.18 μm CMOS," in *International Solid-State Circuits Conference*, 2005, pp. 282–283.

[290] Y.-D. Jeon, S.-C. Lee, K.-D. Kim, J.-K. Kwon, and J. Kim, "A 4.7 mW 0.32 mm² 10 b 30 MS/s pipelined ADC without a front-end S/H in 90 nm CMOS," in *International Solid-State Circuits Conference*, 2007, pp. 456–457.

[291] B.-G. Lee, B.-M. Min, G. Manganaro, and J. W. Valvano, "A 14-b 100-MS/s pipelined ADC with a merged SHA and first MDAC," *IEEE International Journal of Solid State Circuits*, vol. 43, no. 12, pp. 2613–2619, 2008.

[292] B.-M. Min, P. Kim, F. W. Bowman III, D. M. Boisvert, and A. J. Aude, "A 69-mW 10-bit 80-MSample/s pipelined CMOS ADC," *IEEE International Journal of Solid State Circuits*, vol. 38, no. 12, pp. 2031–2039, 2003.

[293] M. Yoshioka, M. Kudo, K. Gotoh, and Y. Watanabe, "A 1.8 v 10 b 210 MS/s CMOS pipelined ADC featuring 86 dB SFDR without calibration," in *Custom Integrated Circuits Conference*, 2007, pp. 317–320.

[294] D. Gubbins, B. Lee, P. K. Hanumolu, and U.-K. Moon, "Continuous-time input pipeline ADCs," *IEEE International Journal of Solid State Circuits*, vol. 45, no. 8, pp. 1456–1468, 2010.

[295] C. A. A. Bastiaansen, D. W. J. Groeneveld, H. J. Schouwenaars, and H. A. H. Termeer, "A 10-b 40-MHz 0.8-μm CMOS current-output D/A converter," *IEEE International Journal of Solid State Circuits*, vol. 26, no. 7, pp. 917–921, 1991.

[296] C. Toumazou, J. B. Hughes, and N. C. Battersby, Eds., *Switched-Currents: An Analogue Technique for Digital Technology*. Peter Peregrinus Ltd., on behalf of the IEE, London, UK, 1993.

[297] A. R. Bugeja, B.-S. Song, P. L. Rakers, and S. F. Gillig, "A 14-b, 100-MS/s CMOS DAC designed for spectral performance," *IEEE International Journal of Solid State Circuits*, vol. 34, no. 12, pp. 1719–1732, 1999.

[298] W. Schofield, D. Mercer, and L. St. Onge, "A 16 b 400 MS/s DAC with < −80 dBc IMD to 300 MHz and < −160 dBm/Hz noise power spectral density," in *International Solid-State Circuits Conference*, 2003, pp. 126–127.

[299] M. Clara, W. Klatzer, A. Wiesbauer, and D. Straeussnigg, "A 350 MHz low-OSR ΔΣ current-steering DAC with active termination in 0.13 μm CMOS," in *International Solid-State Circuits Conference*, 2005, pp. 118–119.

[300] B. Nejati and L. Larson, "An area optimized 2.5-V 10-b 200-MS/s 200-μA CMOS DAC," in *Custom Integrated Circuits Conference*, 2006, pp. 161–164.

[301] B. Jewett, J. Liu, and K. Poulton, "A 1.2 GS/s 15-bit DAC for precision signal generation," in *International Solid-State Circuits Conference*, 2005, pp. 110–111.

[302] M.-J. Choe, K.-H. Baek, and M. Teshome, "A 1.6 GS/s 12 b return-to-zero GaAs RF DAC for multiple Nyquist operation," in *International Solid-State Circuits Conference*, 2005, pp. 112–113.

[303] P. Schvan, D. Pollex, and T. Bellingrath, "A 22 GS/s 6 b DAC with integrated digital ramp generator," in *International Solid-State Circuits Conference*, 2005, pp. 122–123.

[304] D. Baranauskas and D. Zelenin, "A 0.36 W 6 b up to 20 GS/s DAC for UWB wave formation," in *International Solid-State Circuits Conference*, 2006, pp. 580–581.

[305] B. Greenley, R. Veith, D.-Y. Chang, and U.-K. Moon, "A low-voltage 10-bit CMOS DAC in 0.01-mm^2 die area," *IEEE Transactions on Circuits and Systems – Part II*, vol. 52, no. 5, pp. 246–250, 2005.

[306] C.-H. Lin and K. Bult, "A 10-b, 500-MSample/s CMOS DAC in 0.6 mm^2," *IEEE International Journal of Solid State Circuits*, vol. 33, no. 12, pp. 1948–1958, 1998.

[307] G. I. Radulov, M. Heydenreich, R. W. van Hofstad, J. A. Hegt, and A. H. M. von Roermund, "Brownian-bridge-based statistical analysis of the DAC INL caused by current mismatch," *IEEE Transactions on Circuits and Systems – Part II*, vol. 54, no. 2, pp. 146–150, 2007.

[308] S. Luschas and H.-S. Lee, "Output impedance requirements for DACs," in *IEEE International Symposium on Circuits and Systems*, vol. 1, 2003, pp. 861–864.

[309] P. Crippa, C. Turchetti, and M. Conti, "A statistical methodology for the design of high-performance CMOS current-steering digital-to-analog converters," *IEEE Transactions on Computer-Aided Design of Integrated Circuits and Systems*, vol. 21, no. 4, pp. 377–394, 2002.

[310] C. S. G. Conroy, W. A. Lane, and M. A. Moran, "Statistical design techniques for D/A converters," *IEEE International Journal of Solid State Circuits*, vol. 24, no. 4, pp. 1118–1128, 1989.

[311] B. Schafferer and R. Adams, "A 3 V CMOS 400 mW 14 b 1.4 GS/s DAC for multi-carrier applications," in *International Solid-State Circuits Conference*, 2004, pp. 360–361.

[312] G. A. M. Van der Plas, J. Vandenbussche, W. Sansen, M. S. J. Steyaert, and G. G. E. Gielen, "A 14-bit intrinsic accuracy q^2 random walk CMOS DAC," *IEEE International Journal of Solid State Circuits*, vol. 34, no. 12, pp. 1708–1718, 1999.

[313] Y. Cong and R. L. Geiger, "Switching sequence optimization for gradient error compensation in thermometer-decoded DAC arrays," *IEEE Transactions on Circuits and Systems – Part II*, vol. 47, no. 7, pp. 585–595, 2000.

[314] M. Vadipour, "Gradient error cancellation and quadratic error reduction in unary and binary D/A converters," *IEEE Transactions on Circuits and Systems – Part II*, vol. 50, no. 12, pp. 1002–1007, 2003.

[315] G. Manganaro and J. Pineda de Gyvez, "A four quadrant S^2I switched-current multiplier," *IEEE Transactions on Circuits and Systems–Part II*, vol. 45, no. 7, pp. 791–799, 1998.

[316] A. R. Bugeja and B.-S. Song, "A self-trimming 14-b, 100-MS/s CMOS DAC," *IEEE International Journal of Solid State Circuits*, vol. 35, no. 12, pp. 1841–1852, 2000.

[317] Q. Huang, P. A. Francese, C. Martelli, and J. Nielsen, "A 200 MS/s 14 b 97 mW DAC in 0.18 μm CMOS," in *International Solid-State Circuits Conference*, 2004, pp. 364–365.

[318] G. Manganaro, "Analog calibration of a current source array at low supply voltage," US Patent 7 161 412, 2007.

[319] G. Manganaro, "Transresistance amplifier," US Patent 7 202 744, 2007.

[320] Y. Tang, J. Briaire, K. Doris *et al.*, "A 14 b 200 MS/s DAC with SFDR > 78 dBc, IM3 < −83 dBc and NSD < −163 dBm/Hz across the whole Nyquist band enabled by dynamic-mismatch mapping," in *VLSI Circuits Conference*, 2010, pp. 151–152.

[321] W.-H. Tseng, C.-W. Fan, and J.-T. Wu, "A 12-bit 1.25-GS/s DAC in 90 nm CMOS with > 70 dB SFDR up to 500 MHz," in *International Solid-State Circuits Conference*, 2011, pp. 192–193.

[322] W.-H. Tseng, J.-T. Wu, and C.-W. Fan, "A CMOS 8-bit 1.6 GS/s DAC with digital random return-to-zero," *IEEE Transactions on Circuits and Systems – Part II*, vol. 58, no. 1, pp. 1–5, 2011.

[323] J. Hyde, T. Humes, C. Diorio, M. Thomas, and M. Figueroa, "A 300-MS/s 14-bit digital-to-analog converter in logic CMOS," *IEEE International Journal of Solid State Circuits*, vol. 38, no. 5, pp. 734–740, 2003.

[324] D. Marche, Y. Savaria, and Y. Gagnon, "Laser fine-tuneable deep-submicrometer CMOS 14-bit DAC," *IEEE Transactions on Circuits and Systems – Part I*, vol. 55, no. 8, pp. 2157–2165, 2008.

[325] D. Mercer and L. Singer, "A 12-b 125 MSPS CMOS D/A designed for spectral performance," in *IEEE International Symposium on Low Power Electronics and Design*, 1996, pp. 243–246.

[326] A. Van den Bosch, M. S. J. Steyaert, and W. Sansen, "SFDR-bandwidth limitations for high speed high resolution current steering CMOS D/A converters," in *International Conference on Electronics, Circuits and Systems*, vol. 3, 1999, pp. 1193–1196.

[327] J. Deveugele and M. S. J. Steyaert, "RF DAC's: Output impedance and distortion," in *RF Circuits: Wide Band, Front-Ends, DAC's, Design Methodology and Verification for RF and Mixed-Signal Systems, Low Power and Low Voltage*, M. Steyaert, A. van Roermund, and J. H. Huijsing, Eds. Kluwer, Dordrecht, the Netherlands, 2006, pp. 45–64.

[328] A. van den Bosch, M. Steyaert, and W. M. Sansen, *Static and Dynamic Performance Limitations for High Speed D/A Converters*. Kluwer, Dordrecht, the Netherlands, 2004.

[329] A. Van den Bosch, M. A. F. Borremans, M. S. J. Steyaert, and W. Sansen, "A 10-b, 1-GSample/s Nyquist current-steering CMOS D/A converter," *IEEE International Journal of Solid State Circuits*, vol. 36, no. 3, pp. 315–324, 2001.

[330] J. Bastos, A. M. Marques, M. S. J. Steyaert, and W. Sansen, "A 12-bit intrinsic accuracy high-speed CMOS DAC," *IEEE International Journal of Solid State Circuits*, vol. 33, no. 12, pp. 1959–1969, 1998.

[331] M. Clara, A. Wiesbauer, and W. Klatzer, "Nonlinear distortion in current-steering D/A-converters due to asymmetrical switching errors," in *IEEE International Symposium on Circuits and Systems*, vol. 1, 2004, pp. 285–289.

[332] G. Manganaro, "Current steering digital to analog converter with improved dynamic linearity," US Patent 7 023 367, 2006.

[333] D. Seo and G. H. McAllister, "A low-spurious low-power 12-bit 160-MS/s DAC in 90-nm CMOS for baseband wireless transmitter," *IEEE International Journal of Solid State Circuits*, vol. 42, no. 3, pp. 486–495, 2007.

[334] K. Doris, J. Briaire, D. Leenaerts, M. Vertregt, and A. van Roermund, "A 12 b 500 MS/s DAC with >70 dB SFDR up to 120 MHz in 0.18 μm CMOS," in *International Solid-State Circuits Conference*, 2005, pp. 116–117.

[335] K. Doris, A. H. M. van Roermund, and D. Leenaerts, *Wide-Bandwidth High Dynamic Range D/A Converters*. Springer-Verlag, Berlin, Germany, 2010.

[336] M. Alioto and G. Palumbo, *Model and Design of Bipolar and MOS Current-Mode Logic: CML, ECL and SCL Digital Circuits*. Kluwer, Dordrecht, the Netherlands, 2010.

[337] B. Schafferer, "Control loop for minimal tailnode excursion of differential switches," US Patent 6 774 683, 2004.

[338] S. Park, G. Kim, S.-C. Park, and W. Kim, "A digital-to-analog converter based on differential-quad switching," *IEEE International Journal of Solid State Circuits*, vol. 37, no. 10, pp. 1335–1338, 2002.

[339] B. Schafferer, "Constant switching for signal processing," US Patent 6 842 132, 2005.

[340] S. H. Hall, G. W. Hall, and J. A. McCall, *High-Speed Digital System Design: A Handbook of Interconnect Theory and Design Practices*. John Wiley & Sons, New York, NY, 2000.

[341] T. Chen and G. G. E. Gielen, "The analysis and improvement of a current-steering DACs dynamic SFDR – I: The cell-dependent delay differences," *IEEE Transactions on Circuits and Systems – Part I*, vol. 53, no. 1, pp. 3–15, 2006.

[342] T. Chen and G. G. E. Gielen, "The analysis and improvement of a current-steering DAC's dynamic SFDR – II: The output-dependent delay differences," *IEEE Transactions on Circuits and Systems – Part I*, vol. 54, no. 2, pp. 268–279, 2007.

[343] D. Mercer, "Latch with data jitter free clock load," US Patent 7 023 255, 2006.

[344] D. Mercer, "A low power current steering digital to analog converter in 0.18 micron CMOS," in *IEEE International Symposium on Low Power Electronics and Design*, 2005, pp. 72–77.

[345] A. R. Bugeja, "High dynamic linearity current-mode digital-to-analog converter architecture," US Patent 6 906 652, 2005.

[346] K. Gulati, M. S. Peng, A. Pulincherry *et al.*, "A highly integrated CMOS analog baseband transceiver with 180 MSPS 13-bit pipelined CMOS ADC and dual 12-bit DACs," *IEEE International Journal of Solid State Circuits*, vol. 41, no. 8, pp. 1856–1866, 2006.

[347] J. Deveugele and M. S. J. Steyaert, "A 10-b 250 MS/s binary-weighted current steering DAC," *IEEE International Journal of Solid State Circuits*, vol. 41, no. 2, pp. 320–329, 2006.

[348] X. Wu, P. Palmers, and M. S. J. Steyaert, "A 130 nm CMOS 6-bit full Nyquist 3 GS/s DAC," *IEEE International Journal of Solid State Circuits*, vol. 43, no. 11, pp. 2396–2403, 2008.

[349] R. W. Adams and K. Q. Nguyen, "Dual return-to-zero pulse encoding in a DAC output stage," US Patent 6 061 010, 1997.

[350] R. Adams, K. Q. Nguyen, and K. Sweetland, "A 113-dB SNR oversampling DAC with segmented noise-shaped scrambling," *IEEE International Journal of Solid State Circuits*, vol. 33, no. 12, pp. 1871–1878, 1998.

[351] K. O'Sullivan, C. Gorman, M. Hennessy, and V. Callaghan, "A 12-bit 320-MSample/s current-steering CMOS D/A converter in 0.44 mm^2," *IEEE International Journal of Solid State Circuits*, vol. 39, no. 7, pp. 1064–1072, 2004.

[352] X. Wu, P. Palmers, and M. S. J. Steyaert, "An 80-MHz 8-bit CMOS D/A converter," *IEEE International Journal of Solid State Circuits*, vol. 21, no. 6, pp. 983–988, 1986.

[353] T. W. Kwan, R. W. Adams, and R. Libert, "A stereo multibit sigma delta DAC with asynchronous master-clock interface," *IEEE International Journal of Solid State Circuits*, vol. 31, no. 12, pp. 1881–1887, 1996.

[354] K. Nguyen, A. Bandyopadhyay, R. Adams, K. Sweetland, and P. Baginski, "A 108 dB SNR 1.1 mW oversampling audio DAC with a three-level DEM technique," *IEEE International Journal of Solid State Circuits*, vol. 43, no. 12, pp. 2592–2600, 2008.

[355] P. Rombouts and L. Weyten, "Dynamic element matching for pipelined A/D conversion," in *IEEE International Conference on Electronics, Circuits, and Systems*, vol. 2, 1998, pp. 315–318.

[356] K. L. Chan, J. Zhu, and I. Galton, "Dynamic element matching to prevent nonlinear distortion from pulse-shape mismatches in high-resolution DACs," *IEEE International Journal of Solid State Circuits*, vol. 43, no. 9, pp. 2067–2078, 2008.

[357] T. Shui, R. Schreier, and F. Hudson, "Modified mismatch-shaping for continuous-time delta–sigma modulators," in *Custom Integrated Circuits Conference*, 1998, pp. 225–228.

[358] T. Shui, R. Schreier, and F. Hudson, "Mismatch shaping for a current-mode multibit delta–sigma DAC," *IEEE International Journal of Solid State Circuits*, vol. 34, no. 3, pp. 331–338, 1999.

[359] A. Fishov, E. Siragusa, J. Web, E. Fogleman, and I. Galton, "Segmented mismatch shaping D/A conversion," in *IEEE International Symposium on Circuits and Systems*, vol. 4, 2002, pp. 679–682.

[360] B. Nordick, C. Petrie, and Y. Cheng, "Dynamic element matching techniques for delta–sigma ADCs with large internal quantizers," in *IEEE International Symposium on Circuits and Systems*, vol. 1, 2004, pp. 653–656.

[361] Y. Cheng, C. Petrie, B. Nordick, and D. Comer, "Multibit delta–sigma modulator with two-step quantization and segmented DAC," *IEEE Transactions on Circuits and Systems – Part II*, vol. 53, no. 9, pp. 848–852, 2006.

[362] K. L. Chan and I. Galton, "A 14 b 100 MS/s DAC with fully segmented dynamic element matching," in *International Solid-State Circuits Conference*, 2006, pp. 582–583.

[363] K. L. Chan, J. Zhu, and I. Galton, "A 150 MS/s 14-bit segmented DEM DAC with greater than 83 dB of SFDR across the Nyquist band," in *VLSI Circuits Conference*, 2007, pp. 200–201.

[364] K. L. Chan, N. Rakulijc, and I. Galton, "Segmented dynamic element matching for high-resolution digital-to-analog conversion," *IEEE Transactions on Circuits and Systems – Part I*, vol. 55, no. 11, pp. 3383–3392, 2008.

[365] T. Lin and H. Samueli, "A 200-MHz CMOS $x/\sin(x)$ digital filter for compensating D/A converter frequency response distortion," *IEEE International Journal of Solid State Circuits*, vol. 26, pp. 1278–1285, 1991.

[366] A. K. Gupta, J. Venkataraman, and O. M. Collins, "Measurement and reduction of ISI in high-dynamic-range 1-bit signal generation," *IEEE Transactions on Circuits and Systems – Part I*, vol. 55, no. 11, pp. 3593–3606, 2008.

[367] Y. Li and B. Schafferer, "Mixer/DAC chip and method," International Patent WO 2008/112 348 A1, 2008.

[368] Y. Li and B. Schafferer, "Mixer/DAC chip and method," US Patent 7 796 971, 2010.

[369] P. T. M. van Zeijl and M. Collados, "On the attenuation of DAC aliases through multiphase clocking," *IEEE Transactions on Circuits and Systems – Part II*, vol. 56, no. 3, pp. 190–194, 2009.

[370] Y. Zhou and J. Yuan, "A 10-bit wide-band CMOS direct digital RF amplitude modulator," *IEEE International Journal of Solid State Circuits*, vol. 38, no. 7, pp. 1182–1188, 2003.

[371] J. Deveugele, P. Palmers, and M. S. J. Steyaert, "Parallel-path digital-to-analog converters for Nyquist signal generation," *IEEE International Journal of Solid State Circuits*, vol. 39, no. 7, pp. 1073–1082, 2004.

[372] J. Franca, A. Petraglia, and S. K. Mitra, "Multirate analog-digital systems for signal processing and conversion," *Proceedings of the IEEE*, vol. 85, no. 2, pp. 242–262, 1997.

[373] Analog Devices Staff, "A technical tutorial on digital signal synthesis," Analog Devices, Application Note, 1999.

[374] K. Lee, Q. Meng, T. Sugimoto *et al.*, "A 0.8 V, 2.6 mW, 88 dB dual-channel audio delta–sigma D/A converter with headphone driver," *IEEE International Journal of Solid State Circuits*, vol. 44, no. 3, pp. 916–927, 2009.

[375] V. Colonna, M. Annovazzi, G. Boarin *et al.*, "A 0.22-mm^2 7.25-mW per-channel audio stereo-DAC with 97-dB DR and 39-dB SNR$_{out}$," *IEEE International Journal of Solid State Circuits*, vol. 40, no. 7, pp. 1491–1498, 2005.

[376] K. Nguyen and R. Schreier, "System and method for tri-level logic data shuffling for oversampling data conversion," US Patent 7 079 063, 2006.

[377] A. Bandyopadhyay, M. Determan, S. Kim, and K. Nguyen, "A 120 dB SNR, 100 dB THD+N, 21.5 mW/channel multibit continuous time $\Delta\Sigma$ DAC," in *International Solid-State Circuits Conference*, 2011, pp. 482–483.

[378] R. Hezar, L. Risbo, H. Kiper *et al.*, "A 110 dB SNR and 0.5 mW current-steering audio DAC implemented in 45 nm CMOS," in *International Solid-State Circuits Conference*, 2010, pp. 304–307.

[379] L. Risbo, R. Hezar, B. Kelleci, H. Kiper, and M. Fares, "A 108 dB DR, 120 dB THD and 0.5 V rms output audio DAC with inter-symbol-interference shaping algorithm in 45 nm," in *International Solid-State Circuits Conference*, 2011, pp. 484–485.

[380] J. Vankka, *Digital Synthesizers and Transmitters for Software Radio*, 1st edn. Springer-Verlag, Berlin, Germany, July 2005.

[381] H. T. Nicholas III and H. Samueli, "A 150-MHz direct digital frequency synthesizer in 1.25-μm CMOS with –90-dBc spurious performance," *IEEE International Journal of Solid State Circuits*, vol. 26, no. 12, pp. 1959–1969, 1991.

[382] D. De Caro, N. Petra, and A. G. M. Strollo, "A 380 MHz direct digital synthesizer/mixer with hybrid CORDIC architecture in 0.25 μm CMOS," *IEEE International Journal of Solid State Circuits*, vol. 42, no. 1, pp. 151–160, 2007.

[383] C. Y. Kang and E. E. Swartzlander Jr., "Digit-pipelined direct digital frequency synthesis based on differential CORDIC," *IEEE Transactions on Circuits and Systems – Part I*, vol. 53, no. 5, pp. 1035–1044, 2006.

[384] A. Ashrafi, R. Adhami, and A. Milenković, "A direct digital frequency synthesizer based on the quasi-linear interpolation method," *IEEE Transactions on Circuits and Systems – Part I*, vol. 57, no. 4, pp. 863–872, 2010.

[385] D. De Caro and A. G. M. Strollo, "High-performance direct digital frequency synthesizers in 0.25 μm CMOS using dual-slope approximation," *IEEE International Journal of Solid State Circuits*, vol. 40, no. 11, pp. 2220–2227, 2005.

[386] F. F. Dai, W. Ni, S. Yin, and R. C. Jaeger, "A direct digital frequency synthesizer with fourth-order phase domain $\Delta\Sigma$ noise shaper and 12-bit current-steering DAC," *IEEE International Journal of Solid State Circuits*, vol. 41, no. 4, pp. 839–850, 2006.

[387] A. McEwan and S. Collins, "Direct digital-frequency synthesis by analog interpolation," *IEEE Transactions on Circuits and Systems – Part II*, vol. 53, no. 11, pp. 1294–1298, 2006.

[388] S. Thuries, E. Tournier, A. Cathelin, S. Godet, and J. Graffeuil, "A 6-GHz low-power BiCMOS SiGe:C 0.25 μm direct digital synthesizer," *IEEE Microwave and Wireless Components Letters*, vol. 18, no. 1, pp. 46–48, 2008.

[389] S. Mortezapour and E. K. F. Lee, "Design of low-power ROM-less direct digital frequency synthesizer using nonlinear digital-to-analog converter," *IEEE International Journal of Solid State Circuits*, vol. 34, no. 10, pp. 1350–1359, 1999.

[390] J. Jiang and E. K. F. Lee, "A low-power segmented nonlinear DAC-based direct digital frequency synthesizer," *IEEE International Journal of Solid State Circuits*, vol. 37, no. 10, pp. 1326–1330, 2002.

[391] S. E. Turner and D. E. Kotecki, "Direct digital synthesizer with sine-weighted DAC at 32-GHz clock frequency in InP DHBT technology," *IEEE International Journal of Solid State Circuits*, vol. 41, no. 10, pp. 2284–2290, 2006.

[392] Z. Zhou and G. S. L. Rue, "A 12-bit nonlinear DAC for direct digital frequency synthesis," *IEEE Transactions on Circuits and Systems – Part I*, vol. 55, no. 9, pp. 2459–2468, 2008.

[393] X. Geng, F. F. Dai, J. D. Irwin, and R. C. Jaeger, "An 11-bit 8.6 GHz direct digital synthesizer MMIC with 10-bit segmented sine-weighted DAC," *IEEE International Journal of Solid State Circuits*, vol. 45, no. 2, pp. 300–313, 2010.

[394] H. C. Yeoh, J.-H. Jung, Y.-H. Jung, and K.-H. Baek, "A 1.3-GHz 350-mW hybrid direct digital frequency synthesizer in 90-nm CMOS," *IEEE International Journal of Solid State Circuits*, vol. 45, no. 9, pp. 1845–1855, 2010.

[395] J. Savoj, A. Abbasfar, A. Amirkhany, M. Jeeradit, and B. W. Garlepp, "A 12-GS/s phase-calibrated CMOS digital-to-analog converter for backplane communications," *IEEE International Journal of Solid State Circuits*, vol. 43, no. 5, pp. 1207–1216, 2008.

[396] S. Luschas, R. Schreier, and H.-S. Lee, "Radio frequency digital-to-analog converter," *IEEE International Journal of Solid State Circuits*, vol. 39, no. 9, pp. 1462–1467, 2004.

[397] S. Luschas and H.-S. Lee, "High-speed $\Delta\Sigma$ modulators with reduced timing jitter sensitivity," *IEEE Transactions on Circuits and Systems – Part II*, vol. 49, no. 11, pp. 712–720, 2002.

[398] P. Eloranta and P. Seppinen, "Direct-digital RF modulator IC in 0.13 µm CMOS for wideband multi-radio applications," in *International Solid-State Circuits Conference*, 2005, pp. 532–533.

[399] A. Jerng and C. G. Sodini, "A wideband $\Delta\Sigma$ digital-RF modulator for high data rate transmitters," *IEEE International Journal of Solid State Circuits*, vol. 42, no. 8, pp. 1710–1722, 2007.

[400] B. Razavi, *RF Microelectronics*. Prentice Hall, New York, NY, 1998.

[401] P. Eloranta, P. Seppinen, S. Kallioinen, T. Saarela, and A. Pärssinen, "A multimode transmitter in 0.13 µm CMOS using direct-digital RF modulator," *IEEE International Journal of Solid State Circuits*, vol. 42, no. 12, pp. 2774–2784, 2007.

[402] N. Zimmermann, R. Negra, and S. Heinen, "Design of an RF-DAC in 65 nm CMOS for multistandard, multimode transmitters," in *IEEE International Symposium on Radio-Frequency Integration Technology*, 2009, pp. 343–346.

[403] S. M. Taleie, T. Copani, B. Bakkaloglu, and S. Kiaei, "A linear Σ–Δ digital IF to RF DAC transmitter with embedded mixer," *IEEE Transactions on Microwave Theory and Techniques*, vol. 56, no. 5, pp. 1059–1068, 2008.

[404] M. Park, M. H. Perrott, and R. B. Staszewski, "A time-domain resolution improvement of an RF-DAC," *IEEE Transactions on Circuits and Systems – Part II*, vol. 57, no. 7, pp. 517–521, 2010.

[405] B. E. Jonsson, "A survey of A/D-converter performance evolution," in *IEEE International Conference on Electronics, Circuits, and Systems*, 2010, pp. 768–771.

[406] T. Sundström, B. Murmann, and C. Svensson, "Power dissipation bounds for high-speed Nyquist analog-to-digital converters," *IEEE Transactions on Circuits and Systems – Part I*, vol. 56, no. 3, pp. 509–518, 2009.

[407] J.-Y. Wu, R. Subramoniam, Z. Zhang *et al.*, "Multi-bit sigma delta ADC with reduced feedback levels, extended dynamic range and increased tolerance for analog imperfections," in *Custom Integrated Circuits Conference*, 2007, pp. 77–80.

[408] B. J. Farahani, S. G. Krishna, S. Venkatesan *et al.*, "Wide-temperature high-resolution integrated data acquisition for spectroscopy in space," in *IEEE Aerospace Conference*, 2011.

[409] A. Sheikholeslami, R. Payne, and J. Lin, "ISSCC 2009 evening session on 'will ADCs overtake binary frontends in backplane signaling?'," in *International Solid-State Circuits Conference*, IEEE, 2 2009.

[410] E.-H. Chen and C.-K. K. Yang, "ADC-based serial I/O receivers," *IEEE Transactions on Circuits and Systems – Part I*, vol. 57, no. 9, pp. 2248–2258, 2010.

[411] A. Wegener, "Compression of medical sensor data [exploratory DSP]," *IEEE Signal Processing Magazine*, vol. 27, no. 4, pp. 125–130, 2010.

[412] P. Löwenborg, H. Johansson, and L. Wanhammar, "Two-channel digital and hybrid analog/digital multirate filter banks with very low-complexity analysis or synthesis filters," *IEEE Transactions on Circuits and Systems – Part II*, vol. 50, no. 7, pp. 355–367, 2003.

[413] H. Johansson and P. Löwenborg, "A least-squares filter design technique for the compensation of frequency response mismatch errors in time-interleaved A/D converters," *IEEE Transactions on Circuits and Systems – Part II*, vol. 55, no. 11, pp. 1154–1158, 2008.

[414] W. Xiong, U. Zschieschang, H. Klauk, and B. Murmann, "A 3V 6b successive-approximation ADC using complementary organic thin-film transistors on glass," in *International Solid-State Circuits Conference*, 2010, pp. 134–135.

[415] H. Marien, M. Steyaert, N. van Aerle, and P. Heremans, "An analog organic first-order ct $\Delta\Sigma$ ADC on a flexible plastic substrate with 26.5 dB precision," in *International Solid-State Circuits Conference*, 2010, pp. 136–137.

[416] T. Zaki, F. Ante, U. Zschieschang, J. Butschke, F. Letzkus, H. Richter, H. Klauk, and J. N. Burghartz, "A 3.3 V 6b 100 kS/s current-steering D/A converter using organic thin-film transistors on glass," in *International Solid-State Circuits Conference*, 2011, pp. 324–325.

Index